Proof
We're All One
Why Race Is Only Skin Deep

by

Stephen Hawley Martin

WWW.OAKLEAPRESS.COM

Proof We're All One: Why Race Is Only Skin Deep © 2023 by Stephen Hawley Martin. All rights reserved. No part of this book may be used or reproduced in any manner whatsoever without written permission except in the case of brief quotations embodied in critical articles and reviews. For information visit:

www.oakleapress.com

If you find this book to be informative and worthwhile, here are more books by the author you should also consider reading:

Afterlife, The Whole Truth
A Two Volume Set Containing
Life After Death, Powerful Evidence You Will Never Die
and the Sequel, *Heaven Hell & You*

*Life After Death Book Two,
Achieve Joy Now & Bliss in the Afterlife*

*Edgar Cayce, The Meaning of Life
and What to Do About It: Second Edition*

*Your Guide to Achieve Fourth Density
The Law of One, RA, and the End of Suffering*

TABLE OF CONTENTS

Chapter One: We Are All One Life 5
   Darwin's Theory Needs to Be Updated
   The Mind and Quantum Mechanics
   Mind and Matter

Chapter Two: Evolution Is No Accident 26

Chapter Three: Brains Don't Create Intelligence 36
   Deathbed recovery of lost consciousness
   Complex consciousness among people who have minimal brain tissue
   Near Death Experiences
   The Elements of Near Death Experiences
   What about Hell?

Chapter Four: How Many Times Have You Lived? 60
   Sixty Years of Irrefutable Research
   Reincarnation and Christianity
   The Child Who Is His Grandfather
   The Reason Sam May Have Come Back
   A Murder Victim Comes Back
   What Many Cases Have in Common
   Birthmarks and Birth Defects Provide Evidence
   Twenty-Two Percent of UVA Cases Have Birth Defects
   Behavior Traits Also Provide Evidence
   Phobias May Originate in a Former Life
   UVA's Cases May Not Be Representative of the Whole
   Children Aren't the Only Ones Who Have Past Life Memories
   The Past Lives of Glenn Ford

Contents

Chapter Five: Your Beliefs Create Your Reality 69
Chapter Six: Attitude Is Everything 82
    Adopt the right mindset
    How Your Mind Creates Your Life
    Think Positive and Rid Yourself of Fear
    Be Likable and Appealing to Others
    Obey the Laws of Physics
    Always Keep Your Life in Balance
    Above All, Take Time to Make a Life
More Books by The Author 107

Chapter One
We Are All One Life

Growing up in the South during the Jim Crow era, I witnessed blatant discrimination and racism firsthand. It was clearly unfair and unjust, but as a white male immersed in the culture that surrounded me in my hometown—Richmond, Virginia, the capital of the Confederacy—I am sorry to say that until I became an adult and began thinking for myself, I believed the white race to be superior to others. That is how I know what underlies and perpetuates racist thinking, and having begun thinking for myself at the age of 25, why I believe racism can and would end if only the truth were taught in school. Once it becomes widely known that we are all one race under our skin—the human race—racism cannot possibly survive. As you will see as you read ahead, the often unconsciously held belief that some races are inherently superior to others does not hold up to scrutiny.

It seems clear that politics and political action cannot and will not end racism. Legislation was passed 60 years ago that had its elimination as a goal, and it hasn't worked. In 1964, the Civil Rights Act went into effect prohibiting discrimination on the basis of race, color, religion, sex or national origin. The Act not only prohibited discrimination in public accommodations and federally funded programs, it also strengthened the enforcement of voting rights and the desegregation of schools. It seems fair to say that the ideal of those who supported the Civil Rights Act was for society to achieve "color blindness," as articulated by Martin Luther King, Jr. in this famous "I Have a Dream" speech, which took place during the March on Washington on August 28, 1963. Dr. King said, "I look to a day when people will not be judged by the color of their skin, but by the content of their character."

Obviously, it didn't and hasn't worked.

What I believe perpetuates racism may be good intentions in the form of the constant harping by some about "white supremacy" and the "victimhood" of people of color. This keeps people thinking that Caucasians have an inherent advantage and that those without white skin are at a disadvantage. This creates a vicious cycle because reality mirrors people's beliefs. Ask any psychologist. Beliefs are self-fulfilling. If you and others think you have an advantage, at the very least you do in fact have a psychological advantage. It's the same if you and others think you are a victim or in some way inferior. You're going to be at a psychological disadvantage to say the least. What will change this is the widespread belief that everyone is in fact equal, and that under the skin we are all the same as this book will show. When everyone believes that's true, it will be taken for granted and become a non issue.

To make the case, I will cite facts that demonstrate: 1) that beliefs create reality, and 2) that nineteenth century Scientific Materialism and the Darwinian theory that evolution is the result of random mutations and "survival of the fittest," which is still taught in schools and universities as legitimate science, is outdated, incorrect, and serves to perpetuate the erroneous belief that some races are superior to others. 3) That we really are equal because at a deep level we are all one—one life that shares the same consciousness. We are each a facet of a single mind.

Following the presentation of evidence supporting those three facts, I will offer advice to anyone willing to listen that I believe can enable an individual to achieve the best possible life for him or herself.

So let's begin.

Darwin's Theory Needs to Be Updated

Darwinian theory that evolution works solely by random mutations and survival of the fittest lends support to the idea that some human races are more highly evolved than others. It was the underpinning of the imperialist belief in the nineteenth and twentieth centuries of "The White Man's Burden." On top of that, Scientific Materialists, who tend to be fervent supporters of Darwinian theory, would have us believe that humans are analogous to robots with computer-like brains, which naturally leads to the underlying assumption some models are better than others, just as many believe a Cadillac is superior to a Chevrolet. Such thinking led to Jim Crow laws in the American South and to the extermination of six million Jews by the Nazis in World War Two. Once science based on the false assumption that only matter exists is replaced with science that holds up to scrutiny—an overview of which you will soon read—the belief that some races of humans are inherently inferior or superior to others will dissipate and eventually disappear because it is quite simply not true.

As a long-time student of metaphysics and a freethinking Rosicrucian Adept, I believe Jesus was enlightened and knew what he was talking about. Facts are coming that support this contention, and so if you happen to be an atheist, I hope you will not be offended when I quote a few lines that he reportedly said.

In Chapter Ten of the Gospel of John, Jesus explained that it wasn't he but the "Father," i.e., what Christians assume was God working through him that caused the miracles with which he was credited. As part of his explanation he said, "I and the Father are one." (See John 10:30 NIV.) This got him into hot water, and he was about to be stoned by irate Jews.

Jesus replied to the angry mob by saying, "I have shown you many good works from the Father. For which of these do you stone me?" (John 10:32 NIV)

The Jews answered, "We are not stoning you for any good work, but for blasphemy, because you, who are a man, declare yourself to be God." (John 10:33 NIV)

Jesus then quoted Psalm 82:6: "Is it not written in your Law: 'I have said you are gods'?" (John 10:34 NIV)

By quoting this Scripture Jesus clearly was indicating that the Jews who wanted to stone him were "gods," as was he and every other human being—and that includes you—whether you are a Christian or a scientist, black, white, or brown.

By the way, I invite Christians who may scoff at this, as well as those who may think I am the one guilty of blasphemy, to consider these words also spoken by Jesus: "Whoever believes in me will do the works I have been doing, and they will do even greater things than these . . . " (See John 14:12 NIV) To believe in Jesus means, among other things, to believe what he said and taught is true, and that requires believing that we must have the ability, as he said in the Scripture just quoted, to perform miracles.

What I want to communicate is something mystics throughout history have understood: That All-Is-One, and your consciousness, my consciousness, and indeed everyone's consciousness is the single, unified, underlying "I AM" consciousness that Jesus called his "Father." It is the Infinite Mind that underlies, supports, informs—and indeed creates physical reality. You don't have to be religious to understand that—you just have to open your eyes and your mind. It is what Jesus understood, and that understanding and belief is what gave him the power to work miracles. When that has been taught in school for a couple of generations and everyone realizes it, racism will disappear.

I can almost hear someone reading this thinking, "That's a lot of bull."

Really? You doubt that All-Is-One? If so, you ought to read Gary Zukav's book, *The Dancing Wu Li Masters*. In it, he explained quantum mechanics without using complicated mathematics. In summary, any quantum physicist will tell you that matter as envisioned by Newtonian physics does not actually exist. Everything is energy and everything is connected. Consider, for example, the following paragraph from that Gary Zukav's book:

> ... the philosophical implication of quantum mechanics is that all of the things in our universe (including us) that appear to exist independently are actually parts of one all-encompassing organic pattern, and that no parts of that pattern are ever really separate from it or from each other.

Alan Watts [1915-1973], a twentieth century philosopher and interpreter of Zen Buddhism, answered children's questions concerning why they were here, where the universe came from, where people go when they die and so forth with a parable about God playing hide and seek. Watts told them God enjoys the game, but has no one outside himself to play with since he is All-that-Is. He overcomes this problem of not having any playmates by pretending he is not himself. Instead he pretends that he is me and you and all the other people and the animals and rocks and stars and planets and plants and in doing so has wonderful and wondrous adventures. These adventures are like dreams because when he awakes, they disappear. Watts wrote:

> Now when God plays hide and pretends that he is you and I, he does it so well that it takes him a long time to remember where and how he hid himself. But that's the whole fun of it—just what he wanted to do. He doesn't want to find himself too quickly, for that would spoil the game. That is why it is so difficult for you and me to find out

that we are God in disguise, pretending not to be himself. But when the game has gone on long enough, all of us will wake up, stop pretending, and remember that we are all one single Self—the God who is all that there is and who lives for ever and ever.

It will no doubt be shocking to some to think of themselves as God, but Watts was talking about the core essence that is beyond the ego and deeper within than the personal unconscious, the collective unconscious, the archetypes and so on. As Joseph Campbell [1904-1987] said in the PBS TV series, *The Power of Myth*, "You see, there are two ways of thinking 'I am God.' If you think, 'I here, in my physical presence and in my temporal character, am God,' then you are mad and have short-circuited the experience. You are God, not in your ego, but in your deepest being, where you are at one with the non dual transcendent."

As indicated by the above text, that we are all One Life that arises from the "non dual transcendent" is not something I came up with. It is something mystics and enlightened beings have known for thousands of years. But if you believe what you were taught in school or college by your science teachers that nothing exists except matter and that life and evolution happened by accident, you believe, as mentioned above, that the body can be compared to a machine and the brain to a computer. On the other hand, if you are a Christian, you likely believe you were created by God and that when your body dies your spirit will go to heaven—even though no scientific theory I'm aware of supports that view.

No matter what you now believe, I urge you to stick with me, and please keep an open mind as you read ahead. I'm going to relate a theory that I developed after more than forty years of research and study, and I will cite facts to back it up. Not only does it explain what caused the Big Bang and how life on earth came to be, it makes clear that faith is not required to believe we are all one and your consciousness will continue after your body dies. It also indicates that you possess what likely has been until now undreamed of power and that we are spiritual beings having a physical experience. There is no room for racism in any of that.

Two mysteries have captivated the human imagination for thousands of years. The first is why the universe exists at all. Why is there something rather than nothing? The second is that conscious minds exist to perceive it. An ancient idea is that the mystery of consciousness and the mystery of existence are intimately connected, and perhaps surprisingly, there are now a growing number of philosophers and scientists who take this possibility seriously.

Beginning in the first half of the twentieth century, cosmologists began learning a great deal about the early universe by analyzing cosmic background radiation and other phenomena. Using powerful telescopes they were able to see that there are many galaxies, and due to their shift toward the red end of the spectrum of light, that those farthest away are moving away from us faster than those in closer proximity. As a result, cosmologists are able to peer deeply into the past and infer the state of the universe in what is thought to be its first fractions of a second. But where did it all come from? What existed before the Big Bang?

Physicists have proposed that the spark of existence had its origin in a quantum fluctuation, triggering an explosive chain reaction, leading to the still evolving universe we inhabit today. This narrative, however, presupposes the laws of quantum mechanics. As British Biochemist Rupert Sheldrake said in a TED Talk, "[Scientists today say] give us one free miracle and we'll explain the rest.' And the one free miracle is the appearance of all the matter and energy of the universe, and all the laws that govern it, from nothing in a single instant." Suffice it to say that rather than explaining existence, current scientific theories of the origins of the universe have simply pushed things back to a point that raises the question, "What existed before the Big Bang?" Could it all have come from nothing? Although that is apparently what some scientists believe, it doesn't make sense. As the song in *The Sound of Music* goes, "Nothing comes from nothing, nothing ever could."

Instead of beginning with nothing, it seems logical that the challenge of explaining existence should focus instead on defining a self-existing ground of being for which no explanation is required. Physicists have proposed that the true ground floor of reality is the seething quantum realm of particles, forming in and out of existence. While this level of reality surely exists, there is no clear reason why the primordial situation should be constrained by quantum physics. A deeper level of explanation seems to be required, and one possibility is that consciousness is the ground of being. How seething quantum particles or "strings" came to be the ground of reality calls out for an explanation, but in theory, consciousness can explain itself. A unique feature of consciousness is that it does not appear grounded in anything beyond itself. The conscious self is self-producing insofar that it exists only in and to itself. As René Descartes [1596-1650] famously said, "I think therefore I am."

Moreover, without consciousness, it would not matter if anything existed because no sentient being would perceive it. I am reminded of a question that my philosophy professor once asked on a test, "If a tree falls in the woods and no one and no thing were there to hear it, would it make a sound?" The answer is "No. For there to be a sound, someone or something has to hear it." You can say the same about physical reality. Would physical reality exist if no one and no thing were here to perceive it? The same logic says, "No." In a few paragraphs we will see that "observation" is what creates physical reality.

I intuit someone reading this saying to him or herself, "Yeah, but the brain creates consciousness."

Not so. As we will see, the conclusion drawn after sixty years of research at the University of Virginia School of Medicine is that the brain does not create consciousness. The brain is a receiver of consciousness like a cell phone—a receiver that integrates consciousness with the body.

My theory is based on the idea that consciousness just is. Your personal consciousness seems to belong to you because you have memories stored in your unconscious mind. You also have a name and perhaps a job and a history that was created while inhabiting a physical body, and that gives you a sense of identity. When your consciousness leaves your body, you will continue to have those memories—if not indefinitely, at least for a while, according to studies you will soon read about. Even if those memories dissipate and your consciousness eventually dissolves into your higher self and ultimately into the universal consciousness, that universal consciousness will still be you. I say this because consciousness is the "I AM," the Silent Observer at the back of your mind.

The Mind and Quantum Mechanics

Let's take a look at a quantum mechanics experiment that I believe supports the assertion that the minds of humans and the Infinite Mind not only are connected, but are at one with one another. It has to do with how light behaves, and it's called the "Double Slit Experiment." The strange but revealing phenomenon associated with it is known as "The Participating Observer." The findings of this experiment are straightforward, and have been replicated in many laboratories. No honest scientist will refute them.

Scientists have known for more than a hundred years that light can behave both as waves and as particles (photons), but until 1905 they thought light was composed only of waves. Thomas Young (1773-1829) demonstrated in 1803 that light is waves by placing a screen with two parallel slits between a source of light—sunlight coming through a hole in a screen—and a wall. Each slit could be covered with a piece of cloth. These slits were razor thin, not as wide as the wavelength of the light. When waves of any kind pass through an opening not as wide as they are, the waves diffract. This was the case with one slit open. A fuzzy circle of light appeared on the wall.

When both slits were uncovered, alternating bands of light and darkness appeared, the center band being the brightest. Scientists call this a zebra pattern. The areas of light and dark result from what is known in wave mechanics as interference. Waves overlap and reinforce each other in some places and in others they cancel each other out. The bands of light on the wall indicated where one wave crest overlapped another crest. The dark areas showed where a crest and a trough met and canceled each other out.

In 1905, Albert Einstein published a paper that revealed light also behaves as if it consists of particles. He did so by using the photoelectric effect. When light hits the surface of a metal, it jars electrons loose from the atoms in the metal and sends them flying off as though struck by tiny billiard balls. So, Thomas Young demonstrated light is waves, and Einstein demonstrated it is particles. This is the sort of paradox that led scientists to develop quantum mechanics.

Now let's consider a double slit experiment constructed to determine what happens when those conducting the experiment observe or do not observe which slits the photons of light pass through. This time a gun is used that fires one photon at a time. I first read about this experiment years ago in an article entitled, "Faster Than What?" in the June 19, 1995 issue of *Newsweek*. It reported on a paper to be published by a well-known quantum physicist, Raymond Chiao, then teaching at the University of California at Berkeley. Just so you'll know I'm not making this up, in July 2008 I reviewed the facts of this experiment as they are presented here with a guest on the weekly podcast I hosted at the time, the noted quantum physicist Henry P. Stapp, author of *MINDFUL UNIVERSE: Quantum Mechanics and the Participating Observer* (Springer, 2007). He indicated I understood the facts correctly.

Both slits were open and a detector was used to determine which slit a photon passed through. A record was made of where each one hit. Only one photon at a time was shot, so one would suppose there could be no interference. This was the case. The photons did not make a zebra pattern. Rather, they made marks, tiny dots, on a screen.

When the detector was turned off, however, and it was not known which slit a photon passed through, the zebra pattern appeared. In other words, without the detector making it possible for the researcher to observe which slit particles passed through, the particles behave like waves even though they were fired one at a time.

Imagine the stir this caused among those conducting the experiment. In the *Newsweek* article reporting on this, Nobel Prize winning physicist Richard Feynman (1918-1988) was quoted as saying this is the "central mystery" quantum mechanics, that something as intangible as knowledge—in this case, which slit a photon went through—changes something as concrete as a pattern on a screen.

But how could knowledge change the behavior of particles shot from a gun? Materialist science cannot produce an explanation because a tenet of Scientific Materialism is that the brain produces consciousness, awareness, and thought, and that means consciousness, awareness, and thought are confined within a person's skull. Since it would be ludicrous to suggest that thought enclosed inside a person's head could be capable of having an effect on photons shot from a gun, it ought to be clear to everyone that the tenet is false. Obviously, consciousness is not confined inside the skull. Yet Materialists tenaciously cling to the tenet, saying there must be two different sets of laws of physics: a small (subatomic world) set, and a macro world (the one we live in) set. Somewhere between these two worlds, the laws of physics must change.

This hypothesis doesn't explain why thought contained in someone's head should change things at the subatomic level, or anywhere else. Second, other experiments refute the contention that two different laws of physics exist. One such experiment involves large (carbon 60) molecules called "buckyballs," so big they can be seen and therefore are part of the macro world. Research shows they exhibit the same quantum properties as infinitely small particles. Another is an experiment conducted in 2002 that involved crystals. It produced quantum ridges half an inch high—

large enough to be measured with a conventional macro-world ruler.

What makes more sense is what William of Ockham (c. 1287–1347) is thought to have been the first to say, "The simplest explanation is the best." The researcher's ability to know—his or her consciousness—causes the waves to collapse into particles that form a pattern. This happens in my view because there is really only one mind, the Infinite Mind we all share. Therefore, when the researcher—whose mind is at one with the One Mind—can access the knowledge, the zebra pattern does not occur. If he or she cannot access it, the zebra pattern appears. This was verified by setting up the experiment several ways. In the first, the detectors were in front of the two slits. In the second, researchers placed detectors between the screen and the two slits, i.e., after the photons had passed through them. As in the original experiment, knowing about a photon's behavior at the two slits made the zebra pattern vanish, whether or not the detectors were before or after the slits (see the accompanying graphic). But when the detectors were switched off, the zebra stripes returned.

In a third variation, a detector was placed before the slits and a mechanism erased the knowledge after the photon had passed through. The same thing happened. The zebra pattern returned. The result was the same no matter which way the experiment was set up—before the slits, after the slits, or before the slits and then erased. Whether or not the researcher was able to know where each photon hit determined the presence of the zebra pattern, or the

Proof We're All One

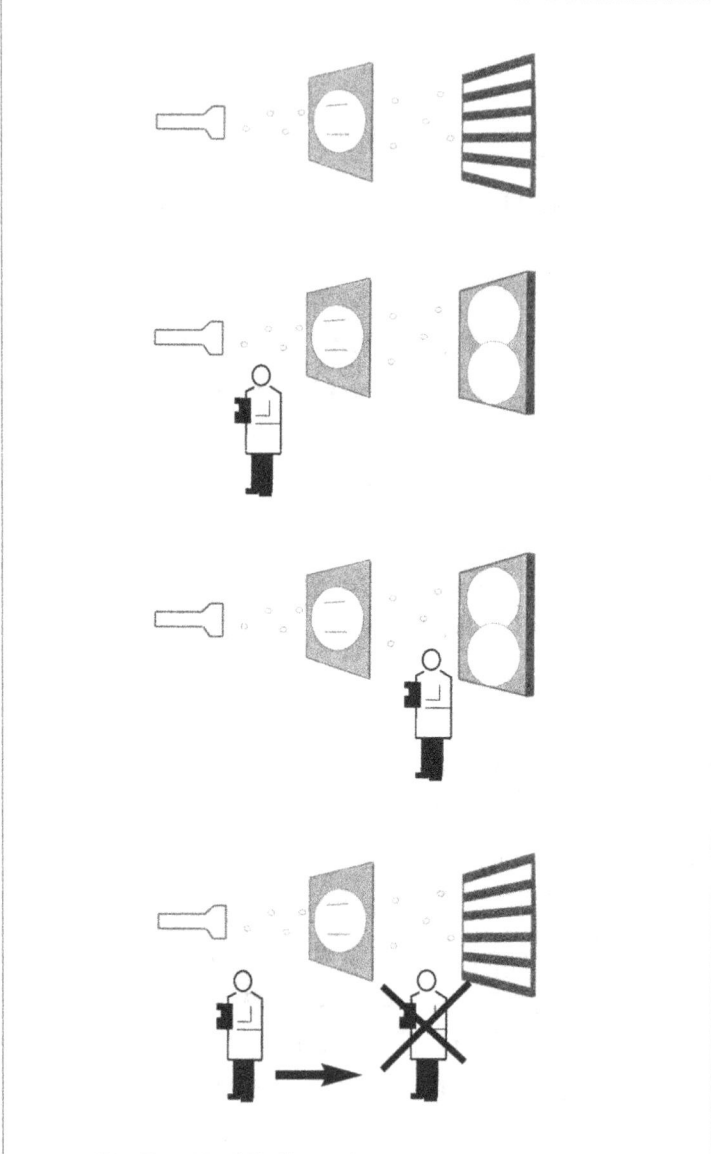

The Double Slit Experiment demonstrates the observer's mind is at one with the Infinite Mind.

lack of it. Versions of the experiment reported on in *Newsweek* were carried out at the University of Munich and at the University of Maryland. The behavior of the photons, the researchers report, "is changed by how we are going to look at them."

It seems impossible to deny that in these experiments consciousness is interacting with physical reality. Moreover, the results demonstrate that future events—what scientists are going to do—can affect earlier events, which indicates that somehow time can double back on itself. This may be because at the subatomic level, time does not exist—time being a phenomenon, some would say a dimension, of physical reality. As you may know, the physicist Stephen Hawking (1942-2018) held that the beginning of the universe was itself the beginning of time.

Since, at the subatomic level, what researchers are going to do in the future causes something to happen in the past, it must also be possible that the birth of the universe could have been caused by something that happened in the universe's future. This is not a new idea. It was originally proposed by the physicist John Wheeler (1911-2008) and later developed by the physicist Paul Davies.

Wheeler pointed out that observation of a quantum system not only defines the state of the system in that moment, but it also defines its entire history. It is as though the act of measurement produces that which is consistent with the present choice to observe. As Wheeler once famously remarked, "We decide what the photon shall have done after it has already done it."

Proof We're All One

In recent decades, this surprising fact of quantum physics has been demonstrated many times in laboratories, and as Wheeler has argued there is no reason to doubt that this principle is true of the entire universe—that observations made now, or perhaps even in the distant future, stretch all the way back to the beginning of the universe, and thereby establish the necessary conditions in which observers can exist.

I realize it is not easy to wrap one's mind around the idea that observations made today, or in the future, caused the Big Bang and perhaps other amazing things such as DNA, and therefore life, but it makes sense in light of the fact that future observations cause non-physical waves to become things known as protons. That everything past, present, and future exists at once is not as far-fetched as it may seem. For the weekly podcast I hosted, I once interviewed the man, F. Holmes Atwater, who set up and between 1978 and 1988 ran the unit for U. S. Army intelligence called the Stargate Project that spied on the Soviet Union and Eastern Block countries using remote viewing performed by psychics. He told me that one of the problems he frequently encountered was determining when an observation that one of his viewers made took place. Apparently, time does not exist in the mental realm beyond matter. Therefore, what a remote viewer saw might be something that took place in the past, or it could be something that will take place in the future. One thing that had to be done was to find a way to determine when what was seen took place.

Mind and Matter

I can almost hear a few readers raising the question, "In the experiments discussed above, mind or consciousness was shown to affect physical reality, but can it create physical reality?"

I am about to provide evidence that indicates it can, evidence that was collected by a college professor named Stephen Braude, whom I also interviewed for the aforementioned weekly podcast. At the time, Dr. Braude was a tenured professor of philosophy at the University of Maryland Baltimore County, and I had just read his book, *The Gold Leaf Lady and Other Parapsychological Investigations* (The University of Chicago Press, 2007).

The interview did not disappoint. Dr. Braude related several well documented and amazing stories of mind over matter, but perhaps the most fantastic, as well as the one that supports the contention that mind [Infinite Mind] creates matter, had to do with Katie, a woman born in Tennessee, the tenth of twelve children.

Katie is apparently a simple woman. Illiterate, at the time Dr. Braude wrote the book about her, she lived in Florida with her husband and worked as a domestic. She was also a psychic who'd had documented successes helping the police solve crimes. In one instance she was able to describe the details of a case so thoroughly and accurately, the police regarded her as a suspect until those actually responsible were apprehended. She apparently also was able to apport objects—in other words, she somehow caused them to disappear in one place and reappear in another, at least that is what Dr. Braude maintained when I spoke with him. And that wasn't all. Seeds reportedly germinated rapidly in her cupped hands. Observers claim to have seen her bend metal, and she was both a healer and a medium or channel. Being illiterate, she could not read or write in her native English, but she has been video taped writing quatrains in medieval French similar both in style and content to the quatrains of Nostradamus.

I know some scientists are going to flip out when they read what comes next because it goes against what they see as a fixed law of physics—that matter cannot be created nor destroyed—but most amazing, perhaps, is what appeared spontaneously on her skin—on her hands, face, arms, legs, and back—apparently out of thin air. It looked like gold leaf, a thin version of the wrapping on a Hersey's Kiss. Katie could not control when this happened, but Dr. Braude and other witnesses saw the foil materialize firsthand. He even videotaped it appearing on her skin.

I just stopped typing and checked. As of this writing, footage from this video can be seen on YouTube. Go to YouTube and put "gold leaf lady Braude" in the YouTube search bar. The title of the video is "UMBC In the Loop: Stephen Braude."

Dr. Braude took the foil to be analyzed. It turned out not to be gold at all, but brass—approximately 80 percent copper and 20 percent zinc.

Dr. Braude thinks there's a reason she produces brass and not gold. Where does the brass foil that appears on Katie's skin come from? It appears that her mind creates it. In fact, as mentioned, Dr. Braude believes she produces brass rather than gold for a reason. You see, Katie had a difficult and tense relationship with her husband. Once she apported a carving set. It just appeared. And her husband—apparently nonplussed—said, "So what? It's not worth anything." Soon afterward, gold colored foil began appearing on Katie's skin. But it wasn't real gold, it was fool's gold—brass. Dr. Braude thinks this is how she gets back at her husband. Katie's mind—albeit the unconscious part—creates matter in the form of brass foil. This being the case, why should it be difficult to believe that an Infinite Mind—one infinitely more powerful than a human mind—created the material universe? The physical universe had to come from something.

One thing quantum physicists agree on is that the physical universe isn't really physical. It is energy—vibrations—in other words, waves. In the double slit experiment, the researcher's ability to know turned waves into photons, which is to say that thought or mind turned waves into things [particles]. It does not seem to me too big a leap to suggest that mind is what creates physical reality, especially once it has been observed.

Chapter Two
Evolution Is No Accident

If matter is all there is, consciousness and intelligence could not have existed until evolution produced a brain. And if evolution really does happen solely as a result of random mutations and "survival of the fittest," then humans of one race could well be more advanced than those of another because their brains have become more evolved. In other words, it would certainly seem possible the brains of the members of some races—perhaps because they lived in a climate that made life less of a challenge—might be less evolved and closer to that of apes—a belief that was commonplace not long ago and no doubt is still held by some today. Of course, most would typically keep such thoughts to themselves because it would be politically incorrect to say something like that out loud.

It follows that if consciousness and intelligence cannot exist without a brain, then consciousness must be located and confined within the gray matter inside the human skull. We know from the previous chapter that isn't so, which raises the question, "How did that mistaken belief come to be?" It's a Materialist view that took hold after Darwin published *The Origin of Species* in 1859. It was a time when science still had a long way to go before it evolved out of a primitive state. The knowledge of how things worked back then was so far behind what it is today that scientists didn't even know that bacteria cause disease. Bacteria were thought to be a symptom rather than a cause. Back then, scientists also thought the atoms that form matter were comparable to tiny marbles—rather than packets of rapidly vibrating units of energy now known as electrons and quarks. Life was believed to have come about by accident when lightning struck a primeval lagoon made up of the right combination of chemicals. Amazingly—and this is something I hope this book will change—that's still what scientists purport to think today. Apparently, those who know the truth either keep quiet about it for fear of being ridiculed by their peers, or they do so because they think it might lead to a resurgence of religion, which they sincerely hope will simply go away.

A lot happened and was discovered between 1859 and 1957 that increased our knowledge—some of which will be discussed—but for now, let's fast-forward to that date. It's the year Francis Crick [1916-2004] discovered that the chemical subunits along the interior of the double helix of DNA function like alphabetic characters in a written language, or the digital characters such as the zeros and ones in a computer code. No doubt you've seen DNA code printouts. Crick realized they direct the construction of proteins and protein machines that all cells need to stay alive. In other words, it came to light that digital information directs the construction of the crucial components of living cells. Therefore, to explain the origin of life, one would have to explain how this complicated processing system came about.

How complicated is it? According to an article on the website of BBC Science Focus Magazine, the UK's leading science and technology monthly: "The DNA in your cells is packaged into 46 chromosomes in the nucleus. As well as being a naturally helical molecule, DNA is supercoiled using enzymes so that it takes up less space. If you stretched the DNA in one cell all the way out, it would be about two meters long and all the DNA in all your cells put together would be about twice the diameter of the Solar System."

How incredible is that? The strand of DNA in a single cell is six feet, six inches long. That must contain an awful lot of code when you consider that the size of the characters in the code is microscopic. Think of the enormous amount of information packed into it. How in the world could lightning striking the chemical soup in a primeval lagoon result in a microscopic strand of computer-like code more than six feet long? The answer is, it couldn't.

So that you are able see for yourself that I'm not making this up, here is the URL of the article referenced above:

https://www.sciencefocus.com/the-human-body/how-long-is-your-dna/

It should go without saying that whenever we see information and trace it back to its source, whether it's computer code, a paragraph in a book, or a computer program, there is always an intelligent input that accounts for that information. This indicates, of course, is that intelligence is behind the origin of life, and yet an ardent Scientific Materialist would argue that given infinite time, anything can occur by accident—for example, that a room full of monkeys with typewriters would eventually produce War and Peace or the complete works of Shakespeare with no typos. It would only be a matter of time. The problem is that mathematicians who have worked out the odds of that actually happening say it would take something like infinity in terms of time and an almost infinite number of monkeys and typewriters—in other words, it is virtually, if not totally, impossible.

The truth is that not enough time was available for DNA to have happened by chance. Scientists tell us the universe began with a Big Bang 13.8 billion years ago, and the earth is about 4.5 billion years old. They say life began on earth 3.77 billion years ago, so that leaves 730 million years for six feet, six inches of microscopic code—with no typos—to have come about by chance. According to mathematicians who have worked out the odds of that happening, it simply could not have come about that way. Of course, it's true some scientists argue against the theory that the universe had a beginning. They believe the universe has always existed and that it contracts and expands. But whether it had a beginning or has always existed and contracts and expands, the result would be the same—it got off to a (perhaps new, after an infinite number of previous) start(s) 13.8 billion years ago. This is indicated by a broad range of phenomena, including the abundance of light elements, the cosmic microwave background (CMB), large-scale structure and Hubble's law, i.e., the farther away galaxies are, the faster they are moving away from Earth.

I can almost hear someone reading this book thinking: "Yeah, well, maybe life was brought here by a space alien." Maybe. But where did he or she get it? And who or what created the space alien?

Let me interject here to head off another possible objection to my theory. A book published in 2019, *Metaphysical Experiments: Physics and the Invention of the Universe* by Bjørn Ekeberg, points to recent measurements of the Hubble constant, the rate of universal expansion, that suggest major differences between two independent methods of calculation. Ekeberg says that discrepancies on the expansion rate have implications not simply for calculation but for the validity of cosmology's current standard model. Ekeberg also tells his readers that another recent probe found galaxies inconsistent with the theory of dark matter, which posits this hypothetical substance to be everywhere, but according to the latest measurements, it is not, suggesting the theory needs to be reexamined.

If this is indeed the case, it appears the various cosmological theories now in vogue may need to be revised. However, my theory is that matter had to come from something and that something is consciousness, or Infinite Mind. It is not dependent on the Big Bang or any other creation theory that is based on the premise that matter is all there is. It follows that after the Infinite Mind created matter, it created life, albeit perhaps as a result of future observations—as has been discussed.

Proof We're All One

It seems to me that anyone who thinks deeply about the intricacy of what is required to produce life would have to come to the conclusion that information resembling computer code that directs something complicated to happen must be the product of some sort of intelligence. And yet that would have been impossible if material substance—matter—is all there is. To repeat what has already been stated, if that were the case, intelligence could not have existed until evolution produced a brain. Could six feet of microscopic code happen by accident? Really?

Not every scientist has turned a blind eye to the facts. After Click's discovery, a number of them began to see that there must be some sort of guiding intelligence responsible for the origin of life. I know this from personal experience because I read a book more than forty years ago that put forth that argument. Published in 1975, it refuted the idea that intelligence, consciousness, and awareness, came about as a result of evolution. The book was entitled *Intelligence Came First*. It was produced by a group of well qualified individuals that met monthly to read and discuss material that was compiled and edited by Ernest Lester Smith [1904-1992], a Fellow of the Royal Society—the prestigious scientific academy of the United Kingdom, dedicated to promoting excellence in science.

Intelligence Came First caused quite a bit of controversy when it came out. Among other things, the book put forth the DNA, computer-code-like argument above, and it also noted that throughout the eons of evolution, needs have preceded the organs through which they are fulfilled—

eyes, ears, taste buds, hearts, kidneys, and so forth. Each new organ developed in response to a need, the book's contributors argued, so why would the brain be an exception? The conclusion the book's authors and contributors came to was that intelligence came first, quite able to function in its own realm. As you will see in an upcoming chapter, scientists at the University of Virginia, where research into this topic has been ongoing for 60 years, have come to the same conclusion: consciousness and intelligence do not need a brain to exist.

I still have a copy of *Intelligence Came First,* which has long been forgotten perhaps by everyone except me—because Materialists shouted it down with a vengeance. But think of the intelligence that would be required to design any one of those organs. Could all of them have come about by chance, i.e., random mutations followed by natural selection? Who that's really thought about the complexity of an eye, a liver, or a kidney could possibly think it could have happened by accident? And yet it appears that back in 1975, the scientific community did just that, and apparently some still think that way today—or refuse to admit that they don't.

How can there be any doubt that intelligence—give it whatever name you like, I prefer "Infinite Mind"—created life? It started with DNA and microscopic one-celled animals. Eventually, that first life evolved into complicated organisms with hearts, eyes, ears, taste buds, hearts, kidneys, and brains. It eventually produced the pinnacle of evolution on Earth—the human being. Did the Infinite

Mind do that with a goal in mind? Probably. Whatever the case may be, one thing seems absolutely certain. The Infinite Mind fosters—perhaps "pushes" is closer to the correct word—toward growth, evolution, and harmony. I say "harmony" because a healthy organism is in harmony with itself—dis-ease being the antithesis of harmony.

In 2007, many years after reading that book, I took the opportunity to become the talk show host and producer of a weekly Internet podcast called *The Truth about Life*. For almost three years I read, and over that period of time interviewed, more than a hundred authors engaged in quests for truth. Among them were medical doctors, parapsychologists, metaphysicians, physicists specializing in quantum mechanics, near death survivors, theologians, psychiatrists, psychologists, and all manner of researchers into the true nature of reality. I don't recall any of these cutting-edge individuals who held to a Materialist point of view, except one guest who could not produce any facts to back up his claims. All he could do was say something to the effect that a particular claim "cannot be so because it goes against what science tells us." I found that about as convincing as a statement by a fundamentalist Christian that, "It can't be so because the Bible says otherwise."

Darwin's theory is that mutations happen randomly and that those that help an organism survive long enough to reproduce are passed onto the next generation. That makes sense and is likely a factor in how organisms adapt when their environments change. But how would that lead to an eye, or an ear, or a kidney?

Suppose, for example, that a chance process does result in something that's moving in the right direction for the creation of a kidney—or to back up a bit, to the creation of computer-like code in DNA that makes life possible in the first place? Entropy, i.e., the natural direction of spontaneous change toward disorder, will work against making further progress. In other words, entropy will likely unwind that progress before additional progress can be made.

In conclusion, the chance-based argument is bogus for two reasons: 1) time works against the chemical synthesis of life, and 2) there is a limit to the amount of time for it to have happened. Using a four-dial bike lock as an example, how likely is it that a thief will be able to break the four-digit code? There are 10,000 possible combinations ($10 \times 10 \times 10 \times 10 = 10,000$). The odds will be against it happening unless the thief has enough time to sample more than half (5,000) of the possible combinations. Therefore, when assessing the plausibility of a random search for an informational sequence, it's necessary to assess how many opportunities there are to do so, versus the complexity of the sequence.

Here is the bottom line with respect to the creation of life. It turns out that when scientists that study this sort of thing do the math, the complexity of sequences is so great just for the standard proteins that make life possible to be built, i.e., there are so many combinations that would have to be searched, that there is not enough time for that to have happened—much less for life to have formed and to have evolved to its current state. The Infinite Mind is

Proof We're All One

infinitely intelligent, and we all evolved from it and remain part of and connected to it. Not only are we all One Life, as we will see in the next chapter, consciousness and intelligence are totally independent of the brain, which means the pigmentation of one's skin has nothing to do with an individual's innate intelligence, and that is something we need to begin teaching in schools and colleges.

Chapter Three
Brains Don't
Create Intelligence

Let me extend my apologies to you if you have read my book, *Afterlife, Powerful Evidence You Will Never Die.* In this chapter, I'm going to summarize a lecture I also summarized in that book. It was recorded on video given in India in 2011 by Bruce Greyson, M.D., then The Chester Carlson Professor of Psychiatry and Director of the Division of Perceptual Studies at the University of Virginia whose job it was to study consciousness. Dr. Greyson is now a professor emeritus.

The bottom line takeaway of Dr. Greyson's lecture is that, despite what many scientists think—black, white, and every color in between—we are all spiritual beings having a physical experience. Dr. Greyson presented an abundance of evidence showing that brains do not actually create consciousness or intelligence, the unmistakable implication being that it must originate somewhere else. He does say, however, that this mistaken belief is understandable since evidence does exist that the brain produces consciousness. Consider what happens when a person drinks too much or gets knocked on the head. Also, it's possible to measure electrical activity in the brain during certain kinds of mental tasks and to identify correlations between different areas of the brain and the different activities. We can stimulate different parts of the brain and record what experiences result, and we can remove parts of the brain and observe what the results are on behavior. This suggests that the brain is involved with thinking, perception, and memory, but according to Dr. Greyson, it does not necessarily suggest the brain causes those thoughts, perceptions, and memories. What the measurements actually show are correlations, rather than causation. The truth is that thoughts, perceptions, and memories, actually occur somewhere else and then are received and processed by the brain in a way similar to how a television, cell phone, or radio receiver works.

Western science, Dr. Greyson pointed out, is largely reductionist. It breaks everything down to its component parts, which are much easier to study than the whole, but the component parts do not always act like the whole. The brain is composed of millions of nerve cells or neurons, but a single neuron cannot formulate a thought, cannot feel angry or cold. It appears that brains can think and feel, but brain cells cannot. No one knows how many neurons are needed in order for them to collectively formulate a thought, nor do we know how a collection of neurons can think when a single neuron cannot.

Scientists get around this by saying consciousness is an emergent property of brains, a property that emerges when a large enough mass of brain cells gets together. According to Dr. Greyson, however, saying something is an emergent property is a way of saying it is a mystery that cannot be explained. It is a fact that there is no known mechanism in the brain or anywhere else that can produce non-physical things like thoughts, memories, or perceptions. The materialistic understanding of the world fails to deal with how electrical impulses, or a chemical trigger in the brain, can produce a thought or a feeling, or for that matter, anything the mind does. Despite this, according to Dr. Greyson, most scientists continue to maintain what he labeled, "The nineteenth century, materialist view that the brain in some miraculous way we do not understand produces consciousness." These scientists, he said, "Discount or ignore that consciousness in extreme circumstances can function very well without a brain."

Dr. Greyson noted that the idea that the mind and the brain are separate is what most people believed until a couple of hundred years ago, but in the nineteenth century western world, beginning with the Darwinians, science began exploring the idea that the physical brain might be the source of thoughts and consciousness. Ironically, as one group of scientists attempted to explain consciousness in terms of Newtonian physics, scientists in a different discipline, physics, were forced to move away from Newtonian physics and develop quantum mechanics in order to explain phenomena in which consciousness—what a researcher knows or doesn't know—completely changes the results of certain experiments. It is as though the right hand did not know what the left hand was up to. Incredibly, this remains how things are today.

Dr. Greyson listed a number of examples in his lecture of evidence researchers with the Division of Perceptual

Studies—established in 1967 at the University of Virginia—have collected that demonstrate that consciousness can exist without a brain being involved. It is a testament to the stubbornness of materialist scientists that even though Dr. Greyson and his colleagues have been collecting this data for fifty years, and many papers and books have been written and published revealing a great deal of it, most western scientists are unaware of this evidence. As a result, you will soon have a leg up on many western scientists.

The evidence falls into four categories:

1. Recovery of lost consciousness in the moments or days prior to death among people who have been unconscious for prolonged periods of time.
2. Complex consciousness ability in some people who have minimal brain tissue.
3. Complex consciousness in near-death experiences when the brain is not functioning or is functioning at a greatly diminished level.
4. Memories, particularly among young children, accurately recalling details of a past life.

Deathbed recovery of lost consciousness

The unexpected return of mental clarity shortly before death by patients suffering from neurological or psychiatric disorders has been reported in western medical literature for more than 250 years. There are published cases in the medical literature of patients suffering from brain abscesses, tumors, strokes, meningitis, Alzheimer's disease, schizophrenia, and mood disorders, all of whom long before had lost the ability to think or communicate. In many of these cases, evidence from brain scans or autopsies showed their brains had deteriorated to an irreversible degree, and yet in all of them, mental clarity returned in the last minutes, hours, and sometimes days before the patients' deaths. The Division of Perceptual Studies has identified 83 cases in western medical literature and has collected additional unpublished contemporary accounts wherein patients recovered complete consciousness just before death. It appears as though the damaged brain released its grip on a patient's mind and clarity returned as a result.

In 1844, a German psychiatrist named Julius reported that this occurred in 13 percent of patients who had died in his institution. In a recent investigation of end of life experiences in the United Kingdom, 70 percent of caregivers in nursing homes reported that they had observed patients suffering from dementia and confusion becoming completely lucid in their last hours before death. In a case Dr. Greyson himself investigated, a 42-year-old man developed a malignant brain tumor that rapidly grew in size. He quickly became bedridden, blind in one eye, unable to communicate, incoherent and bizarre in this behavior. He appeared unable to make any sense of his surroundings, and when members of his family touched him, he would slap as though being annoyed by an insect. He eventually stopped sleeping and would talk deliriously throughout the night making no sense. After several weeks of this, he suddenly appeared calm and began speaking coherently. He then slept peacefully. The following morning, he remained completely clear and talked with his wife, discussing his imminent death for the first time. He then stopped speaking and died.

There is no known physiological mechanism to explain this phenomenon. It is rare, but the fact that it happens has no explanation in terms of how the brain functions. It suggests the link between consciousness and the brain is more complex than most scientists think. It is as though the damaged brain prevents the person from communicating, but when the brain finally begins to die, consciousness is released from the degenerating brain.

Complex consciousness among people who have minimal brain tissue

Another phenomenon is the presence of normal or even high intelligence in people who have very little brain tissue. There are rare but surprising cases of people who seem to function normally, with normal intelligence, and normal social function, despite having virtually no brain at all. In one case, published in 2007, a high school honor student who had been accepted for enrollment by Smith College underwent surgery after she was injured and knocked unconscious in an automobile accident. An x-ray of her head just before surgery revealed that she had no cerebral cortex at all. She had just a brainstem inside her skull. When the surgeon opened her skull to operate that is exactly what he found—a brainstem and that's all.

Neurologists tell us the brainstem relays motor and sensory signals to the cerebellum and the spinal cord and integrates heart function, breathing, wakefulness, and animal functions. They also tell us the brainstem does not have the connections to perform higher cognitive functions such as thinking, perceiving, making decisions, and so forth. According to scientific knowledge as it now stands, this college-bound honor student should not have been able to formulate a thought of any kind, let alone function at a high intellectual level.

Hers is not an isolated situation. Dr. Greyson pointed to dozens of cases of patients with hydrocephalus, wherein as much as 95 percent of a brain is incapacitated due to an excess of cerebrospinal fluid, and yet many with that level of affliction have normal and even above average intelligence.

Near Death Experiences

The near death experiences [NDEs] Dr. Greyson covered in the lecture were accounts given by people who had been clinically dead for a short time and then resuscitated or revived spontaneously. He said they typically have memories of vivid sensory imagery, and an extremely clear memory of what they experienced. They often describe what they experienced as seeming "more real" than their everyday life. All of this occurs under conditions of drastically altered brain function under which the materialist model would say is absolutely impossible. Such memories are reported by between ten and twenty percent of those who are revived from clinical death. Dr. Greyson has personally investigated almost one thousand cases.

The average age at the time of the near death in these cases was 31 years, but there was a very wide range. A young girl reported an experience she'd had at eight months old while undergoing kidney surgery. The oldest to experience near death Dr. Greyson has studied was 81 at the time of his heart attack. About one third of the NDEs occurred during surgical operations, a quarter during serious illness, and another quarter as a result of life-threatening accidents. The common features of NDEs can be categorized as changes in thinking, changes in emotional state, as well as paranormal and otherworldly features.

Changes in thinking include a sense of time being altered. Often people report that time stopped or ceased to exist. The change in thinking phenomenon also included a sudden revelation or change in understanding in which everything in the universe suddenly became crystal clear. There was a sense of the person's thoughts going much faster and being much clearer than usual. Finally, there was a life review—a panoramic memory in which the person's life seemed to flash before him or her.

Typical emotions reported included an overwhelming sense of peace and wellbeing, a sense of cosmic unity and of being one with everything, a feeling of complete joy, and a sense of being loved unconditionally.

The paranormal features included a sense of leaving the physical body, sometimes called an out of body experience [OBE], a sense of physical senses such as seeing and hearing becoming more vivid than ever before. Sometimes people report seeing colors and hearing sounds that do not exist in this life, and a sense of extrasensory perception, i.e., of knowing things beyond the normal ability of the senses, such as things that are happening at a remote location. Finally, some report having visions of the future and that they entered another, unearthly world or realm of existence.

Many report they came to a border they could not cross, a point of no return that if they had crossed they would not be able to return to life. Many also say they encountered a mystical or divine being, and some report seeing spirits and loved ones who died previously and seem to be welcoming them into another realm, or in some cases sending them back to life.

As a psychiatrist, the profound after effects of NDEs are of particular interest to Dr. Greyson. Near death survivors reliably report a consistent pattern of changes in attitudes, beliefs, and values, which do not seem to fade over time. They report overwhelmingly they are more spiritual because of their experience, that they have more compassion, a greater desire to help others, a greater appreciation for life as well as a stronger sense of meaning and purpose in life. A large majority reports they have a stronger belief that we survive death of the body and no longer fear death. About half report they have lost interest in material possessions, and many report they no longer have an interest in obtaining personal prestige, status, or in competition.

Dr. Greyson said that three features of NDEs suggest consciousness is not produced by the brain: 1) Enhanced mental function while the brain is incapacitated; 2) Accurate perceptions from outside the body, such as the ability to accurately tell doctors and nurses what they saw and heard going on in the operating room; and 3) encounters with deceased persons who convey accurate information no one else could have known, including in some instances encounters with deceased persons the NDE survivor could not have known were dead at the time.

In one case, a nine-year-old boy with meningitis had an NDE in which he saw several deceased relatives, including his sister who told him he had to return to his body. As soon as he returned from death, he told his parents—who had been at his bedside for 36 hours during his ordeal. His father became very upset because his daughter was at college in a different state and was perfectly healthy as far as the father knew. The boy insisted that his sister had sent him back and had told him she had to remain.

The father left the hospital, promising his wife he would call their daughter as soon as he got home. When he tried to call her, he learned that the college officials had been trying to contact him and his wife all night to tell them the tragic news. Their daughter had been killed in an automobile accident around midnight.

By the way, if you would like to see a video of Dr. Greyson's lecture just summarized, go to YouTube and search "Dr Bruce Greyson consciousness independent of the brain." A video of the lecture should come up at the top of the list.

The Elements of Near Death Experiences

In December, 2008, I interviewed consciousness researcher Jody Long, who along with her husband, Jeffrey P. Long, M.D., founded the Near Death Experience Research Foundation. They maintain what they believe is the largest NDE web site (www.nderf.org) in the world. It has more than 1800 full-text published NDE accounts.

According to Jody, five steps seem to be common to NDEs:

1. A sense of being dead, including the sudden awareness of a fatal accident, or of not surviving an operation.
2. An out-of-body experience; the sensation of peering down on one's body. In many cases, those experiencing clinical death report back the scene with uncanny accuracy, quoting doctors and witnesses verbatim.
3. Some kind of tunnel experience, a sense of moving upward or through a narrow passage.
4. Light, including light "beings," God or a God-like entity. For those having a hell-like experience, the opposite may be true—darkness or a lack of light.
5. A life review—being shown one's life, sometimes highlighting mistakes or omissions.

I find the life review of particular significance, and it never fails to come to mind whenever I'm tempted to do something that potentially might harm another. Here is what Raymond Moody, M.D., author of *Life After Life* and other books on this subject had to say about the life review:

> *When the life review occurs, there are no more physical surroundings. In their place is a full color, three-dimensional, panoramic review of every single thing the [persons having this experience] have done in their lives.*
>
> *This usually takes place in a third-person perspective and doesn't occur in time as we know it. The closest description I've heard of it is that the person's whole life is there at once.*
>
> *In this situation, you not only see every action that you have ever done, but you also immediately perceive the effect of every single one of your actions upon the people in your life.*
>
> *So for instance, if I see myself doing an unloving act, then immediately I am in the consciousness of the person I did that act to, so that I feel their sadness, hurt, and regret.*
>
> *On the other hand, if I do a loving act to someone, then I am immediately in their place and I can feel the kind and happy feelings.*
>
> *Through all of this, the Being is with those people, asking them what good they have done with their lives. He helps them through this review and helps them put all the events of their life in perspective.*
>
> *All of the people who go through this come away believing that the most important thing in their life is love.*

For most of them, the second most important thing in life is knowledge. As they see life scenes in which they are learning things, the Being points out that one of the things they can take with them at death is knowledge. The other is love.

What about Hell?

Not everyone has such a rosy near death experience. Some say they went to hell, and that the life review was a horrid experience. A few years ago, I gave a talk at a Unity Fellowship Church in Williamsburg, Virginia, during which I touched on the life review and its implications. Afterward, a man came up to me and told me about an NDE and life review a friend of his had recounted. I'll call the man's friend Ralph.

Apparently, before his NDE, Ralph had anger issues and a hair-trigger temper. Some time before Ralph's NDE, he had been driving along, minding his own business, when a car pulled out in front of him, causing a fender-bender. Ralph became enraged, jumped out of his car, pulled the offending driver out of his, and beat him to a pulp. This beating left the offending driver permanently handicapped and unable to earn a living. Ralph was charged with assault and punished by the legal system, but that punishment was nothing compared to what he experienced during his life review.

Not only did Ralph feel the hurt and sorrow of the man he'd injured, he also experienced the sorrow and misery of everyone who had been affected, and everyone who would be affected, including the man's children who were unable to attend college and had to go to work at early ages because of the family's reduced income. Ralph felt the hurt and sorrow experienced by the man's wife and other family members who were forced to care for the injured man and to work at menial jobs in order to put food on the table. The list of those whose sorrow Ralph experienced was long. It included the injured man's descendants who would be affected in future generations by the domino effect the beating Ralph gave would eventually cause. As you might suspect, Ralph was deeply affected by this, turned his life around, and became a different man as the result of his NDE and life review.

A similar story comes to mind, which I heard from a man interviewed for my radio show. He told me about a hell-like experience he had while clinically dead. This was not the same as a life review. The man said he had literally gone to hell because of the life he'd led up to that point as the member of a malicious gang. The experience affected him so deeply that he completely changed and became a Christian pastor.

Rather than relate the experience he described, however, I have extracted a passage from an Amazon.com review by a reader of one of my books that contains information about near death experiences. I think this conveys what my radio guest wanted to communicate. It appears to have been written by a nurse:

While reading this book, I was reminded of a patient I had several years ago. She was in the final stages of death, and it was very early in the morning. She was screaming this horrid scream (one of those you just don't forget) and so I went in to sit with her. She had no family with her and I couldn't bear to let this woman breathe her last on her own. After sitting there for a few minutes, she asked me to uncover her feet, that they were really hot. A few minutes later, she asked me to uncover her legs. This went on for a while and soon she was drenched in sweat and writhing in pain. She told me to be a better person than she was because she was experiencing hell and it was something she hoped no one would ever have to experience. Before her final breaths, she was wrapped in ice-covered blankets to hopefully make her transition somewhat more peaceful. This was the most horrid death that I've ever witnessed.

The International Association of Near-Death Studies estimates the incidence of distressing near death experiences ranges from 1% to 15% of all NDEs, which suggests the percentage of those who have a hell-like experience is low. I seriously doubt, however, that Adolph Hitler, Joseph Stalin, or Jeffrey Dahmer went to heaven. Nevertheless, based on thousands of reported Near Death Experiences, it seems likely we will all someday have a life review. That knowledge by itself should be enough to deter a thinking individual from harming others by practicing tribalism or engaging in racist activities.

Chapter Four
How Many Times Have You Lived?

In his lecture, Dr. Greyson also recounted information about the Division of Perceptual Studies' research into children's memories of past lives. Researchers at the University of Virginia have been conducting these investigations for about sixty years and as a result have in excess of 2500 cases in their files. I was quite familiar with this even before I saw Dr. Greyson's lecture because of research I had done for my book, *REINCARNATION: Good News for Open Minded Christians and Other Truth Seekers.* I have in fact twice interviewed one of the Perceptual Division's key researchers who has written two books on the Division's reincarnation research findings, Jim B. Tucker, M.D., a Phi Beta Kappa graduate of the University of North Carolina, a medical doctor and a board certified child psychiatrist who at the time I spoke with him served as medical director of the Child & Family Psychiatry Clinic at the University of Virginia Medical School..

Sixty Years of Irrefutable Research

Suppose you were changing your son's diaper—let's say he was just beginning to talk and was quite verbally adept at the age of 18 months—and he looked you in the eye and said, "When I was your age, I used to change your diaper."

What would you think?

If your father happened to be deceased, would you possibly think your son might be your father reincarnated? That would make him his own grandfather.

Can something like that happen? Dr. Tucker told me that it can.

At the time I spoke with Dr. Tucker, about 1600 of the 2500 cases—many of which came from Dr. Tucker's predecessor, the late Ian Stevenson, M.D. (1918-2007—had been entered into a computer database along with the information collected on each. The data were sorted into about 200 different variables, allowing researchers to comb through and cross tabulate the data to spot trends as well as to categorize and compare the similarities and differences based on various factors and characteristics.

Dr. Stevenson was a methodical and meticulous researcher who graduated first in his medical school class at Canada's McGill University. He never actually claimed reincarnation as fact, but rather, said his cases were "suggestive" of reincarnation. His often-cited first book on the subject was published in 1966 and entitled, *Twenty Cases Suggestive of Reincarnation.*

The cases he studied come from all over the world. When Dr. Stevenson began this research, they were easiest to find in places where people have a belief in reincarnation such as India and Thailand. This may be because parents were not as likely to think a child was imagining a past life, and because they are not likely to be embarrassed to talk about it. Nowadays, however, people in the United

States are not as reticent as they once were. Dr. Tucker said that since the University of Virginia set up a website on this subject some time ago, he and his colleagues hear from parents "all the time" about their children's memories of past lives.

Nevertheless, in the United States reincarnation is thought by many to go against Christian doctrine, even though recent surveys show that more than twenty percent of Christians believe in reincarnation. The percentage is higher, by the way, among younger adults.

Reincarnation and Christianity

I'd like devout Christians who may be reading this to know about a man I interviewed on my radio show in the spring of 2008. His name is James A. Reid Sr., and he's a Southern Baptist minister, now retired. He holds a Doctor of Ministry degree from San Francisco Theological Seminary. For 15 years he was Chaplain to the Los Vegas strip, where he heard a lot of talk about Edgar Cayce and past lives, which he always dismissed as fantasy thinking. Finally, he got so fed up he decided to write a book denouncing reincarnation as a Biblically untenable doctrine. But Dr. Reid is an honest and mature individual. Once he dug into Church history and the Scriptures, he was forced to change his view. He ended up writing a book that maintains the Bible supports the doctrine of reincarnation. It is called, *BORN AGAIN AND AGAIN AND AGAIN: A Bible-Based View of Reincarnation.*

Dr. Reid maintains that for the first five hundred year history of the Church, many accepted reincarnation as fact. It wasn't until 553 A.D. that it was condemned by the Council of Constantinople, and then only by a narrow margin. He gives several examples indicating Jesus and others of his time believed in reincarnation. For example, John the Baptist was supposed by many to be the prophet Elijah reincarnated. Jesus himself said this was so. (See Matthew 11:14.) Once, Jesus asked his followers who people thought he (Jesus) was. They replied that many believed him (Jesus) to be one of the prophets—presumably reincarnated, since the last prophet died about 400 years earlier. Also, consider the story of Jesus restoring the sight of the man who had been born blind, as recounted in John 9:1-12, in which Jesus' disciples ask him if the man's sins caused his blindness, or if the sins of his parents had caused him to be born blind.

Since the man was blind from birth, the only way his own sins could have caused his blindness was for him to have sinned in a former life. Jesus did not tell his followers this wasn't possible. To the contrary, he seems to have assumed it was possible, although he gives another reason for the man's blindness, saying, "Neither this man nor his parents sinned, but this happened so that the work of God might be displayed in his life."

Edgar Cayce, whose psychic readings probably did more than anything to promote the concept of reincarnation in the West, was a devout Presbyterian and Sunday school teacher who read the Bible once through for every year of his life. At first, when reincarnation started showing up in his readings, he was baffled and confused. But he reread the Bible and satisfied himself it wasn't anti-Christian.

There are many references to reincarnation in the Bible but believers overlook or misinterpret them because they have been conditioned to think reincarnation is taboo. Kevin Todeschi, Executive Director of the Association for Research and Enlightenment, said on my radio show in November 2007, that he has counted eleven such references in Matthew's gospel alone.

The Child Who Is His Grandfather

In the case mentioned at the beginning of this chapter—the 18-month-old child who said he had changed his father's diaper when he was his father's age—the child's mother was the daughter of a Southern Baptist preacher. As you might imagine, she found what her son said to be highly unusual. I asked Dr. Tucker to describe the case when he came on my show, and he obliged.

The child's grandfather had died eighteen months before the child's birth. His first mention of having been his own grandfather was during that change of diapers, but as time went by he made more comments about how he used to be big, and what he did when he was. His mother in particular became interested and began to ask the boy, whose name was Sam, questions. Sam came up with some very specific statements. For instance, she asked him if he had had any brothers or sisters. He said he had had a sister who was killed. In fact the grandfather's sister had been murdered sixty years before.

The parents felt certain the child could not have known this since they had only recently learned about it themselves.

The child also talked about how, at the end of his previous life, his wife would make milkshakes for him every day, and that she made them in a food processor rather than in a blender. This turned out to be true.

When Sam was four years old, his grandmother—his wife in his previous life—died. Sam's dad traveled to where she lived and took care of the estate. When he returned, he brought some family photos with him.

One night Sam's mother had the pictures spread out on the coffee table. Sam walked over and pointed to pictures of his grandfather and said, "Hey, that's me."

To test him she pulled out a class photo from the time the grandfather was in elementary school. Sam ran his finger across the photo, which had sixteen boys in it, and stopped on the one who had indeed been his grandfather.

"That's me," he said.

The Reason Sam May Have Come Back

The grandfather may have come back as the son of his own son because of the relationship—or lack thereof—the two had had in his previous life. The grandfather had not had an open relationship with Sam's dad. He had been a very private person. Sam's dad felt that if his father had really returned as his son, his father may have decided to come back to try to develop a closer bond than had existed in their previous relationship. Dr. Tucker said this may be true. When he visited the family he could see that Sam and his dad were very close.

A Murder Victim Comes Back

Another story from the files of the University of Virginia, Dr. Tucker related on my show has to do with an Indian girl named Kum Kum, who said she had been murdered in her previous life—poisoned—by her daughter-in-law. Kum Kum said she was from a city of about 200,000 located about 25 miles away. One of the things that makes this a good case is that her aunt wrote down a number of statements—eighteen in all—she made before an effort was undertaken to see if they checked out.

All of them did.

The statements included the name of a son, the name of a grandson, the fact that the son had worked with a hammer. And a number of other specifics—for example, that she had a sword hanging near the cot where she slept, and a pet snake she fed milk to.

Research led to the woman Kum Kum claimed to have been, who had died five years before she was born. A big family flap had taken place over a will and who would inherit the worldly possessions of the deceased woman's son. Kum Kum had probably been right. Circumstantial evidence indicated the son's wife had poisoned her mother-in-law—the woman Kum Kum insisted she'd been.

What Many Cases Have in Common

These case histories are fascinating and convincing, and we could go on almost indefinitely considering them, individually. After all, there are more than 2500 in UVA's files. Instead, let's step back and look at the overall findings of this exhaustive study.

Children who report past-life memories typically begin talking about a previous life when they are between two and three years old. Those who have studied this phenomenon believe that emotional involvement with past-life family members argues for reincarnation as the cause rather than superpsi at work—the psychic reservoir of memories housed in the Infinite Mind also known as the Akashic Records. The children tend to show strong emotional involvement with such memories and often tearfully ask to be taken to the previous family. Once there, not only is a deceased individual usually identified whose life matched the details given, during the visits, children often recognize family members or friends from that individual's life. Tearful reunions are common.

Birthmarks and Birth Defects Provide Evidence

Many children studied also had birthmarks that matched wounds on the body of the deceased individual. To give one example, a boy in Thailand, who said he'd been a schoolteacher in his previous life, was shot and killed when riding his bicycle to school one day. He gave specific details including his name in that life and where he had lived. He continued to make this claim until his grandmother took him to the previous address. The child was able to identify the various members of his previous family by name.

Even more startling, he was born with two birthmarks: a small round birthmark on the back of his head and a larger, more irregularly shaped one near the front. The woman he claimed was his wife in that life recalled investigators saying her husband had been shot from behind. The investigators said they knew this because he had a typical, small, round entrance wound in the back of his head and a larger, irregular exit wound in front.

In another case, a boy remembered a life in a village not far away in which he had lost the fingers of his right hand in a fodder-chopping machine. The child was born with an intact left hand but the fingers of his right hand were missing.

The average length of time between the death and rebirth of the children in these birthmark cases is only fifteen to sixteen months. It seems to me, this sort of thing may happen when the soul takes a shortcut between lives, skipping a process by which the life just lived would have been fully integrated into the soul.

Twenty-Two Percent of UVA Cases Have Birth Defects

According to Dr. Tucker's book, Life Before Life (St. Martin's Griffin, 2005), about 22 percent of the cases in the University's database include birth defects due to wounds suffered in violent deaths in the previous life. Most of the cases come from the Hindu and Buddhist countries of South Asia, the Shiite peoples of Lebanon and Turkey, the tribes of West Africa, and the tribes of northwestern North America.

In 1997 Stevenson published details of 225 cases in a massive work Reincarnation and Biology: A Contribution to the Etiology of Birthmarks and Birth Defects. The same year he presented a summary of 112 cases in a much shorter book, Where Reincarnation and Biology Intersect.

In many cases postmortem reports, hospital records, or other documents were located and consulted that confirmed the location of the wounds on the deceased person in question matched the birthmarks. These often correspond to bullet wounds or stab wounds, and as in the case described above. Sometimes two marks correspond to the points where a bullet entered and then exited the body.

Birthmarks are also related to a variety of other wounds or marks, not necessarily connected with the previous personality's death, including surgical incisions and blood left on the body when it was cremated. A woman run over by a train that sliced her right leg in two was reborn with her right leg absent from just below the knee. A man born with a severely malformed ear had been resting in a field at twilight, mistaken for a rabbit, and shot in the ear.

Behavior Traits Also Provide Evidence

Further evidence for reincarnation comes from what might be called behavioral memories. For example, cases exist where children of lower caste Indian families believe they had been upper class Brahmins, and in their view still were. These children would refuse to eat their family's food, which they considered polluted. Conversely, a child remembering the life of a street-sweeper—a very low caste—showed an alarming lack of concern about cleanliness. Some children demonstrate skills they have not learned in their present life, but which the previous personality was known to have had. A number of Burmese children who recalled being Japanese soldiers killed there during World War Two preferred Japanese food such as raw or semi-raw fish over the spicy Burmese fare served by their families.

Many children express memories of their previous lives in the games they play. A girl who remembered a previous life as a schoolteacher would assemble her playmates as pupils and instruct them with an imaginary blackboard. A child who remembered the life of a garage mechanic would spend hours under a family sofa "repairing" the car he pretended it to be. One child who remembered a life in which he had committed suicide by hanging himself had the habit of walking around with a piece of rope tied round his neck.

Phobias May Originate in a Former Life

Phobias occur in about a third of the cases and are nearly always related to the mode of death in the previous life. For example, death by drowning may lead to fear of being immersed in water; death from snakebite may lead to a phobia of snakes; a child who remembers a life that ended when he was shot may display a phobia of guns and loud noises. A person who died in a traffic accident may have a phobia of cars, buses, or trucks.

Sexual orientation may also be affected by a previous life. In one of his books, Ian Stevenson wrote, "Such children almost invariably show traits of the sex of the claimed in the previous life. They cross-dress, play the games of the opposite sex, and may otherwise show attitudes characteristic of that sex. As with the phobias, the attachment to the sex and habits of the previous life usually becomes attenuated as the child grows older; but a few of these children remain intransigently fixed to the sex of the previous life, and one has become homosexual."

Certain preferences and cravings can also carry over. They frequently take the form of a desire or demand for particular foods not eaten in the child's present family, or for clothes different from those ordinarily worn by the family members. Other examples include cravings for addictive substances, such as tobacco, alcohol, and other drugs that the previous personality was known to have used.

UVA's Cases May Not Be Representative of the Whole

Dr. Tucker pointed out that the cases he and others have studied may not be typical because most children do not remember past lives. As mentioned, the average time between lives in these cases is only fifteen months or so—although there are outliers that range up to fifty years. In 70 percent of these cases, the previous personality died by unnatural means. Many died young. This may speed up the reincarnation process. The consciousness may come back quickly due to unfinished business, or because he or she feels shortchanged. The quick return may also be the reason past life memories are intact, as well as sexual preferences, cravings and so forth. My guess is that a much longer duration between lives is the norm. Teachings of the Rosicrucians, a mystical order of which I have been a member and attained the rank of "Adept," say the human personality span is normally about 140 years. If we live 70 years, for example, we can expect to spend 70 years in the realm between lives before we incarnate again. If we live 60 years, we can expect to spend 80 years between lives. The teachings stress, however, that this is a rule of thumb. Centuries could elapse between incarnations, or as with many in the UVA study, the return could come in a matter of months.

Children Aren't the Only Ones Who Have Past Life Memories

Memories of past lives also sometimes occur in adults, and such memories can be of lives that took place long ago. I once recalled a romantic interlude from a life as a Russian army officer during the Napoleonic Wars. The memory was triggered when I met someone that appeared to be the same woman. A guest on my show recalled having a spontaneous recollection of a life as a woman that took place in twelfth or thirteenth century France. He was being burned at the stake. He said this was so vivid it seemed more than a memory. He actually felt he was there, subjectively experiencing the ordeal.

He'd been meditating when suddenly it seemed he was back in the skin he'd occupied then—the action taking place around him. Information about who he was and what was taking place was present in his mind as though he had literally been transported back in time and reentered that body. He said that, surprisingly, he did not feel much pain at being burned—his consciousness exited his body as soon as the flames engulfed it. He floated nearby observing his body burn—not feeling any pain at all. Nevertheless, it was a gut wrenching, emotional experience that left him so distraught he secluded himself after reliving the experience and was unable to communicate with others for two or three days.

He said his need to withdraw had not been because of the pain he'd endured. It was a result of the distress he felt over the pain and suffering humanity puts itself through—man's inhumanity to man. He'd been burned at the stake in that life because he'd been a priestess of the Cathar religion, a Gnostic Christian sect persecuted and eventually extinguished by the Roman Catholic Church. His death by fire was just one of many that took place during the twelfth through fourteenth centuries.

Why were these people killed? No doubt they were seen as a threat to those in power, which at that time included leaders of the Church. The Church taught and teaches that salvation comes through belief in Christ and his sacrifice on the cross. Gnostics followed Christ's teachings but believed salvation comes through direct knowledge of God—a direct and personal relationship. Today, many if not all churches foster this direct relationship—a daily walk with God is considered a requisite by most. Catholicism in the time of the Cathars, however, taught that only the clergy could have this direct relationship.

The Past Lives of Glenn Ford

Anyone with an open mind who looks into what has been found will find it difficult to refute that reincarnation can and does happen. There are many anecdotal accounts. If you are old enough, you may remember Glenn Ford (1916-2006), a movie actor who flourished during Hollywood's Golden Age. Once, when he was approached about taking a role in a movie about Dutch psychic Peter Hurkos (1911-1988), Ford decided he ought to learn something about the paranormal. So, at the age of 54, Ford personally viewed demonstrations by Hurkos, conducted a number of interviews with experts on the subject, and in December 1975, voluntarily underwent three past-life regression sessions via hypnosis, during which he described five previous lives. As is typically done in such sessions, the hypnotized subject, Glenn Ford in this case, was regressed back to childhood, and then was coaxed further back before his current birth to recall previous lives.

In one session, Ford described himself as a bachelor music teacher named Charles Stewart of Elgin, Scotland who died in 1892. Stewart loved horses but hated his job teaching music to young schoolgirls. Amazingly, under hypnosis, Ford agreed to demonstrate his musical skill, and played passages from Beethoven, Mozart, and Bach. When Ford listened to the tapes of the interview, he said he shared Stewart's love for horses and had, since his early years, been considered a natural with those animals. Most significantly, however, he said that in his current life he did not, and could not, play the piano. Following the past-life regression sessions, researchers went to Scotland and located historical records of a music teacher named Charles Stewart of Elgin, Scotland who died in 1892.

A second hypnotic regression session with Ford brought out memories of a life as a member of French King Louis XIV's elite horse cavalry. Under hypnosis, Ford not only gave accurate information about his surroundings in France 300 or so years prior, he was able to speak French fluently—although in his current incarnation, he did not know the language.

Subsequently, Ford was regressed to other previous lives, describing a Christian martyr killed by lions in the Coliseum in third century Rome, and a seventeenth Century Royal Navy sailor who died of the Great Plague. In his most recent prior lifetime, Ford was a cowboy who herded cattle in the American West. It is interesting to note that although Ford starred in 106 movies, as well as several TV series ranging from comedy to police dramas to war stories, he was best known for and most often cast as a cowboy in Westerns.

What does this suggest about you? In the millions of years since the early ancestors of modern humans first walked planet earth, you have probably lived dozens of times, perhaps hundreds, and maybe even thousands. If that is indeed the case, and it seems likely to me, you have been evolving, sometimes rapidly, and in some lifetimes, perhaps not so much. You have no doubt played many roles, from loser to winner and from warrior to wise elder. Everything you have experienced and learned is recorded in your subconscious mind—what Christians would call your soul.

Here's something else to realize and internalize: In some lives you have probably had black skin, in others white, and in some it may have been brown. Anthropologists and paleontologists tell us humans first evolved in Africa and that much of human evolution occurred on that continent. Therefore, if you have been around and evolving for quite some time, it's likely your skin hasn't always been the shade or color it is today. The fossils of early humans who lived between six and two million years ago, for example, come entirely from Africa. Most scientists currently recognize some 15 to 20 different species of early humans. This by itself should be sufficient to say that racism and tribalism make no sense at all. Moreover, in some lives you have been a man and in others a woman. If you don't feel comfortable with the sex you came into this life with, perhaps in your most previous past life, you were a member of the opposite sex. If you have unexplained phobias, fears, or predilections—even talents as Glenn Ford had with horses—they very likely stem from events or conditioning that took place in a previous life. Research

Proof We're All One

by UVa indicates this sort of thing is commonplace. Whatever the case may be, everything you have experienced is now part of you in what some would say is your subconscious mind, and what others would call your Soul or Higher Self.

Chapter Five
Your Beliefs Create Your Reality

There's an important reason why understanding that consciousness is the ground of being can be a boon to humanity in general, minority communities in particular, and to you personally. It has to do with the ability of each individual to create the best possible life for him or herself and for his or her family and loved ones. Understanding that your consciousness is not isolated or trapped within your skull, and in fact is connected to and part of the Infinite Mind, will make it possible for you and others to open doors you may not have realized were there.

You may be thinking that if the future can create the past, the future must be fixed and therefore cannot be changed. Everything indeed exists at once, but the future is not fixed. What exists at the subatomic level of mind is an eternal here and now with a potential future based on what has occurred in the past. The premise of the movie series *Back to the Future* is correct. If things continue in the direction they are now headed, the potential future will result because that future actually does exist in the dimension of mind. But we humans have the power of free will, and that means we have the power to change the future with our actions and our thoughts. Change the direction things are headed, and the future will change.

Proof We're All One

Take a look at these words of Jesus: "Therefore I tell you, whatever you ask for in prayer, believe that you have received it, and it will be yours." (Mark 11:24)

Notice the tense change in that verse. Jesus is saying to believe that you already possess or have achieved what you ask for and you will have it in the future. Let me say, parenthetically, that it's also important to experience the emotion, or to feel the joy of having received it, as this will help establish in the subjective mind that underlies, supports and informs reality that whatever you want is a done deal. By doing so, and by acting accordingly, you will have changed the direction things were going. Jesus apparently knew that thoughts are things and that what we believe already exists does indeed already exist in the nonphysical, timeless realm of mind. It exists as a thought form, and thoughts are things that, unless changed or dismissed, will eventually materialize on the physical plane. As long as you believe you are a victim, you will be a victim. Think negative and act negative, and negative events will follow. Believe you have something important to offer society and others, and you will have something to offer society and others. Think positive and act positive, and the chances a positive future will unfold will increase exponentially.

For those who until now have bought hook, line, and sinker into the false premise of Scientific Materialism, it will take time to realize this. To illustrate how our mental framework can blind us, consider what Charles Darwin found on his visit to Micronesia during his voyage on the Beagle. At that time, the natives of those islands were so isolated from the rest of the world that they had never seen a ship. Darwin and others from the Beagle came ashore in dinghies. The natives had no difficulty seeing them. After all, they themselves used small boats. But they did not, apparently could not, see the Beagle moored offshore—even when it was pointed out to them. A boat of its size did not fit into their mental framework. As a result, it was invisible to them. The same is true today when it comes to the true nature of reality. Then, as now, the way that most see the world is what keeps it hidden. I have to tell you that it took a while for me to realize, for example, the power of belief. Looking back, I recall a startling revelation that came to me in the spring of 2000 following a local radio interview about one of my novels. It was evening. I was beat, having just spent an intense hour trying my best to be entertaining and witty. On my way home, I stopped at my local Seven Eleven for a bottle of beer. A sign caught my eye as I approached the register.

"We I.D. under 27 years of age."

I took my place in line behind a couple of teenagers with Slurpees. An acquaintance from college took the spot behind me, and we exchanged pleasantries. My turn came, I put the bottle on the counter and reached for my wallet.

The clerk eyed me. "Sorry, I'll have to see your I.D.," she said.

"Excuse me?" I said.

"I'm going to need to see your I.D.," she repeated.

"You're kidding," I said.

She let out an exasperated sigh. "No, I need to see your I.D. before I can sell you that beer."

I placed my driver's license in her hand, turned to my friend, and gave a little shrug. Her mouth gaped. "It's true," she said, shaking her head. "You really do look young."

On the way home, I sipped, kept an eye out for police, and pondered the fact that I'd been asked to prove I was old enough to buy alcohol. You see, I was fifty-five years old at the time—more than twice what the clerk was required to I.D. It's definitely true that I felt much younger. Even today, twenty years later, I don't feel a great deal different than I felt when I actually was twenty-seven.

After that encounter, I started wondering why I appeared to be so young, and a possibility surfaced in my mind. Thirty years earlier, when I was 25, I'd read an article about a study of people who'd been consuming large doses of vitamin E for ten years. The article said that no measurable signs of aging had occurred among them. So I went out and bought a bottle, and I've been taking it since.

For years, I believed I wouldn't age. And for years, it seemed I didn't age.

Much later, I read that researchers had concluded that vitamin E in pill form cannot be proven to retard aging. As has often been the case, newer studies refute older ones. But I kept taking it anyway.

According to recent articles, we've come almost full circle. No researcher is ready to say vitamin E stops aging altogether, but new research indicates that taking the vitamin results in lower incidence of heart disease and cancer, while helping mitigate all sorts of health problems. Even so, I've come to believe that back then it may have worked for me in large measure due to the placebo effect. But it worked. Thirty years before I had read an article that said I wouldn't age if I took it. I expected it to work, so it did. If the following week I'd read another article that said the anti-aging qualities of vitamin E were hogwash, I probably would not have experienced the same result.

Belief is extremely potent. The effectiveness of placebos has been demonstrated time and again in double blind scientific tests. The placebo effect—the phenomenon of patients feeling better after taking dud pills—is seen throughout the field of medicine. One report says that after thousands of studies, hundreds of millions of prescriptions and tens of billions of dollars in sales, sugar pills are as effective at treating depression as antidepressants such as Prozac, Paxil and Zoloft. What's more, placebos cause profound changes in the same areas of the brain affected by these medicines, according to this research. For anyone who may still have been in doubt, this proves beyond a doubt that thoughts and beliefs can and do produce physical changes in our bodies.

In addition, the same research reports that placebos often outperform the medicines they're up against. For example, in a trial conducted in April, 2002, comparing the herbal remedy St. John's wort to Zoloft, St. John's wort fully cured 24 percent of the depressed people who received it. Zoloft cured 25 percent. But the placebo fully cured 32 percent.

Proof We're All One

Taking what one believes to be real medicine sets up the expectation of results, and what a person expects to happen usually does happen. It has been confirmed, for example, that in cultures where belief exists in voodoo or magic, people will actually die after being cursed by a shaman. Such a curse has no power on an outsider who doesn't believe. The expectation causes the result. If you've read my novel, *The Secret of Life,* you know I used this phenomenon as a factor in the plot.

Let me relate a real-life example of spontaneous healing that concerned a woman I'd known for a dozen years.

Nancy is a minister's wife. She's a devout Christian—as firm a believer in her religion as a bushman who'd drop dead from a witch doctor's curse is in his. Five years ago, a lump more than half an inch in diameter was discovered in one of her breasts. Her doctor scheduled a biopsy.

A prayer group gathered at her home the night before this procedure was to take place. They prayed not that the lump would be benign, but rather, that it would disappear entirely.

Nancy is a member of a denomination that takes the Bible literally. In Matthew 18:19-20, Jesus is reported to have said, "Again, I tell you that if two of you on earth agree about anything you ask for, it will be done for you by my Father in heaven. For where two or three come together in my name, there I am with them."

As you can imagine, it was more than two or three. It was a living room full. As in my case and vitamin E, quite naturally, Nancy expected the prayers to work.

Belief is powerful. It is the key to manifesting your desires. A study carried out on the Discovery TV Channel, for example, gives an indication. In this case, two researchers conducted the same ESP experiment in the same laboratory using the same equipment. They went to great pains to keep everything identical except for one thing. One researcher believed ESP was valid and the other did not. Both tests were supervised by impartial observers, including the Discovery Channel TV crew.

The experiment that employed the researcher who believed in ESP had a statistically significant number of correct scores, indicating the experiment was a success. The validity of ESP was demonstrated scientifically. But the correct hits in the experiment with the doubting Thomas researcher were within parameters that could be accounted for by chance, meaning the experiment failed to demonstrate the validity of ESP. Apparently, the one and only variable—belief—made the difference. The first researcher believed and the second didn't. Each got the result he expected.

The same thing is at work in prayer by believers. In other books I have written, I report on several studies that demonstrate the efficacy of prayer. Prayer works. Prayer is thought released into the subconscious. Prayers give spirit, or the Life Force, extra zest that bolsters its natural tendency to organize matter in a way that is beneficial to life.

Let's get back to Nancy. The next morning, upon self-examination, the lump in her breast appeared to have vanished. Nonetheless, Nancy kept her appointment at the hospital where her doctor conducted a thorough examination.

The lump indeed was gone. Not a trace could be found, and the bewildered doctor sent her home.

How could a solid lump of tissue disappear? It melted away due to the potent combination of belief, prayer, and expectation. We indeed create our own reality. A way that this might happen is explained in lectures I came across years ago by a man named Thomas Troward. He first delivered them at Queens Gate at Edinburgh University in Scotland in 1904. Called *The Edinburgh Lectures on Mental Science,* they provide a plausible explanation that fits with the findings of studies on prayer—that distance between those praying and the one being prayed for is not a factor, and that the one being prayed for does not have to know about the prayers on his or her behalf. How prayer works is simple, but let me lay some groundwork before I place it before you.

It helps to begin by considering the difference that appears to exist between what we think of as "dead" matter and something we recognize as alive. A plant, such as a sunflower, has a quality that sets it apart from a piece of steel. The sunflower will turn toward the sun under its own power. When first picked, it possesses a kind of glow. This quality might be called the Life Force, or consciousness. On the other hand, the piece of steel appears totally inert. Yet, at the quantum level, the steel is alive with motion. As has been said, quantum physicists tell us that motion or energy is what comprises all matter. Atoms and molecules are not solid things. They are energy—vibrations. Some would say the whole universe is alive, as though it were a single giant thought—the thought of an infinitely vast mind of organizing intelligence.

Even so, by outward appearances the sunflower is alive, and the steel is not. Few would argue this. But one might argue that a plant's state of "aliveness" is different from an animal's. Consider the difference in aliveness between a sunflower, an earthworm, and a goldfish. Each appears to be progressively more alive.

Now, let's add a dog, a three year old child, and a stand up comedian on a late-night talk show. Each has a progressively higher level of intelligence. So, to some extent, what we call the degree of "aliveness" can be measured by the amount of awareness or intelligence displayed—in other words, by the power of thought.

As has been written, consciousness, or intelligence, underlies and creates the entire universe. But it becomes more evident to us—we can see it more clearly—as this intelligence becomes more self-aware. So the distinctive quality of spirit, or life, is thought, and the distinctive quality of matter, as in the piece of steel, is form.

Consider for a moment form versus thought. Form implies the occupation of space and also limitation within certain boundaries. Thought (or life) implies neither. When we think of thought or life as existing in any particular form we associate it with the idea of occupying space, so that an elephant may be said to consist of a vastly larger amount of living substance than a mouse. But if we think of life as the fact of "aliveness," or animating spirit, we do not associate it with occupying space. The mouse is just as alive as the elephant, notwithstanding the difference in size. Here is an important point. If we can conceive of anything as not occupying space, or as having no form, it must be present in its totality anywhere and everywhere—that is to say, at every point of space simultaneously.

Life/thought/consciousness not only does not occupy space, it transcends time. In physical reality it takes time for something to move from one point in space to another. So when there is no space there can be no time. In other words, if life/thought/consciousness is devoid of space, it must also be devoid of time. It seems to me that is how the future can cause the past, as in the quantum mechanics experiments, and how believing you already have it will cause it to happen. The implication is that all life, or thought, must exist everywhere at once in a universal here and an everlasting now.

How does this help us understand how we create our own reality as well as how prayer works?

First, it is implicit in the discussion above that there are two kinds of thought. We might call them lower and higher, or subjective and objective because what differentiates the higher from the lower is the recognition of self. The plant, the worm, and perhaps the goldfish possess the lower kind only. They are unaware of self. Perhaps the dog, and certainly the boy and the comedian possess both. The higher variety of self-aware thought is possessed in progressively larger amounts as if ascending a scale. The top level is our consciousness, but all the levels are connected down to the subjective, ground-of-being mind. Here are the levels of mind as related to me by a professor at the College of Metaphysics in Windyville, Missouri:

1. An Individual's Conscious Mind
2. An Individual's Unconscious Mind
3. An Individual's Subconscious Mind or Soul
4. The Collective Subconscious Mind
5. Subjective, Non-Dual Ground-of-Being Mind

Level Five, the ground of being, subjective mind, is the organizing intelligence or mind present everywhere that, among other things, supports and controls the mechanics of life in every species and in every individual. It causes the plant to grow toward the sun and to push its roots into the soil. It causes hearts to beat and lungs to take in air. It controls all of the so-called involuntary functions of the body. And the truth is, it controls a lot more.

Now let's consider an important point made in the lectures. The conscious mind has power over the subjective mind that creates our reality. I discovered the truth of this firsthand in college when I learned to hypnotize others. I would put a willing classmate into a trance and tell him he was a chicken or a dog. Much to the amusement of my audience, he would then act accordingly.

Hypnotism works because the hypnotist bypasses his subject's conscious mind and speaks directly to the subject's subjective mind. This part of the subject's mind has no choice but to bring into reality that which is communicated directly to it as fact by a conscious mind. Being totally subjective, it cannot step outside of itself and take an objective look. As such, it is capable only of deductive reasoning, which is the kind that progresses from a cause (the conscious mind's directive) forward to its ultimate end, having the mind of a golden retriever. It does not stop to question or analyze. This is the reasoning that a criminal might use in committing a crime. He may walk into a room, see a man counting his money, and think: "I need money, so I will take his. Since the man is protecting the money, I will get rid of him. I'll shoot him. He'll drop to the floor. I will then take the money and run. I'll leave by the window." The subjective mind is non-dual. Right and wrong, good and bad, aren't considered. Only how to get to the end result is.

On the other hand, the conscious mind, being objective and self-aware, can step outside. It can reason both deductively and inductively. To reason inductively is to move backward from result to cause. A police detective, for example, would arrive at the crime scene and begin reasoning backward in an attempt to tell how the crime was committed and who might have done it.

The result is that the subjective mind is entirely under the control of the conscious (objective) mind. With utmost fidelity, the subjective will work diligently to support or to bring into reality whatever the conscious mind believes to be true. Since the individual's subjective mind blends with the ground-of-being mind and is present everywhere, it is able to influence circumstances and events so that whatever the conscious mind believes to be true will indeed become true. So, for example, if I believe I am a sickly person, I will be a sickly person. If I believe that by sitting in a draft I will catch a cold, I will catch a cold when I sit in a draft. Conversely, if I believe that I am rich, that I deserve to be so because it is my birthright, I will become rich even if this is not already the case. If I think I am unlucky, I am unlucky.

This also explains how, why and when prayer works. When people who pray sincerely believe their prayers will have a positive effect, their prayers most certainly will. The belief they hold is impressed upon their own subjective minds. Their subjective minds blend into the ground-of-being mind. The more people praying and believing, the greater the effect. The subjective mind of the person for whom they are praying is also part of the ground-of-being mind, and the latter goes to work to bring about positive results.

It does not make any difference whether the person praying is at someone's bedside or halfway around the world. As noted above, thought, and therefore prayer, is present everywhere at once. It is nonlocal. This explains why prayer is not hindered by distance.

Most people go through life hypnotized into thinking that they have little or no control over their circumstances. The fact is that they create their circumstances with their thoughts and beliefs. The message of the *Edinburgh Lectures* is simple. Change your beliefs and your circumstances will change. And while you are at it, a few well-intentioned prayers can't hurt.

Chapter Six
Attitude Is Everything

We know from sixty years of research by scientists at the University of Virginia that we are not physical beings. Rather, we are spiritual beings having a physical experience, and the human body is simply a vehicle. Just as automobiles come in red, green, white and black, human bodies—our vehicles in physical reality—come in different colors. When it comes to other cars on the road, what ought to be important is not the paint, but how well the person who is behind the wheel drives his or her car. It follows, as Martin Luther King Jr. famously said, that humans should not be "judged by color of their skin but by the content of their character."

That being the case, this chapter imparts information intended to be useful to anyone and everyone who would like to hone their driving skills in an effort to cruise into a better life. Assuming you are willing and have the desire, whether you are black, white, brown or some other shade or color, there is no doubt in my mind you have what it takes to make the best one possible for yourself.

Adopt the right mindset

Some people, the so-called "privileged," are scripted from birth for success. I suggest they should think of themselves not so much as "privileged," but rather, as "fortunate," and devote their energy to doing worthwhile work for their fellow humans in particular, and to society in general. They were fortunate their parents were wealthy, and they were surrounded from their very first breath by wealth and the advantages that they can afford. For them, a wealthy mindset is second nature. If, however, you do not have a wealthy mindset, you need to develop one. You see, in my experience, what's in a person's mind is critically important when it comes to achieving success. Of course, it's true rich kids typically have practical advantages beyond a wealth-mindset that disadvantaged kids do not, such as the wealthy people they know, and therefore the connections they have. Perhaps they went to more prestigious schools. Maybe they also have easier access to the capital needed to start a business. All that may be true, but the advice in this chapter is meant to help you overcome a lack of connections and capital by explaining how to become a "people magnet" that draws people to you who want to join in what your are doing and help you succeed. In order for that to happen, one thing you must do is use your willpower to adjust your thinking and your beliefs so that they mirror and perhaps even improve upon what is in the minds of those so-called "privileged" individuals.

You may be wondering why you need to change your thinking. The answer is simple. Just ask any trained therapist. You focus upon things and take actions based on your thoughts and attitudes, and what you focus on expands. If you think of yourself as a victim of circumstances, you will most certainly be a victim of circumstances. If you think of yourself as wealthy—that you deserve to be rich because it is your birthright—with a sufficient amount of effort, you will become wealthy. That is, in fact, the basic message of two of the most popular self-help, get-rich books of all time: *Think, and Grow Rich* by Napoleon Hill [1883-1970] and *The Science of Getting Rich* by Wallace D. Wattles [1860-1911].

Give this some thought. You and billionaire Warren Buffett were born with the same basic equipment: a mind with huge potential. Both of you are descended from a long line that goes all the way back to the very first single-cell creature that formed in the primordial sea. Both of you are the products of evolution that took place over a mind-boggling 4.5 billion years. Other hominids that evolved along the way branched off onto dead-end paths or developed into chimpanzees or gorillas and such. Some, like the Neanderthals, made it pretty far along the evolutionary path, but eventually could no longer hack it and became extinct when your ancestors, Homo sapiens, came along and took over their territory. But your line kept going and going. Your ancestors continued evolving until they eventually arrived at the very top of the food chain. There can be no doubt at all. You are a member of a very exclusive club—you are one among the most gifted and highly intelligent creatures that has ever lived. You are the pinnacle of life on earth.

Now consider this. Your mind is fantastic. Scientists say we humans typically use only a portion of its capacity. It is the most important tool you have—an amazing tool that is very much like a garden. You can cultivate it, pull out the weeds, water it, plant the right seeds, and allow them to grow. Or you can neglect it, and let it run wild. Either way, cultivated or not, it must and it will bring forth whatever is allowed to grow in it. If no good seeds are planted in your mind and allowed to flourish, useless, destructive weeds will take over and will continue producing more of their kind. Just as a gardener or farmer cultivates his plot of land, weeding it, and growing the flowers, the fruits and the vegetables he or she wants on the dinner table, so may a person tend his or her mind, weeding out the useless and destructive thoughts and cultivating only those that have the promise of bearing delicious fruit. If such cultivation does not take place, if discipline is not exercised, the result likely will not be good because what's in your mind will eventually be what's in your world.

You may be saying to yourself that you cannot help what thoughts enter your mind, and that may be true. But you can decide which thoughts to keep and which thoughts to throw away. Over time, the outer conditions of a person's life always come to be in tune with his or her inner state. By the process of planting and cultivating positive, constructive thoughts, you will sooner or later discover that you are the master gardener of your mind—the director of your life. You will also come to understand that your thoughts shape your character, which also creates your circumstances. Ultimately, what's in your mind is your destiny. A long time ago a man named James Allen [1864-1912] wrote a book about this that I highly recommend you read called, *As a Man Thinketh.*

Proof We're All One

How Your Mind Creates Your Life

Perhaps you are not convinced that what I have just written above is true. How does what's in your mind create your life? I'll explain. As you no doubt already know, the human mind has two parts: a conscious part and a subconscious part. You have direct control over the conscious part. At present, you are causing it to read these words. The reason it is vitally important to use your willpower to control and cultivate your conscious mind is that it has power over your subconscious mind, and to a great extent, your subconscious mind is what determines your circumstances and your reality. Not only does your subconscious mind blend into the mind we all share and influence events either favorably or unfavorably based on how your conscious mind has programmed it, as you go about your daily routine, your subconscious mind influences the decisions you make. As mentioned before, if you have a victim mentality, your subconscious mind may dismiss or filter out all sorts of opportunities that come your way because it determines what you notice and are attracted to out of the literally millions of things you are exposed to each day. If in your subconscious you are convinced you are a victim and nothing you do can change that, your subconscious mind will likely dismiss out of hand and cause you to miss all sorts of opportunities that might lead to success if you only noticed them and consciously and took advantage of them.

How will changing your conscious mindset from victimhood to "destined-to-be-wealth-off –and-secure-because-it-is-your-birthright" change this? Well, consider the phenomenon of hypnotism. The reason it works is because the hypnotist bypasses the subject's conscious mind and speaks directly to the subject's subconscious mind, and the subconscious mind has no choice but to bring about that which is communicated directly to it as fact. Why? The subconscious mind is totally subjective, meaning it cannot step outside itself and take an objective look at what is going on. As such, it is capable only of deductive reasoning, which is the kind that progresses from a cause (the conscious mind's directive) forward to its ultimate end. It does not stop to question or analyze, or think about what is "good," "bad," or "desirable." This is the sort of reasoning a sociopathic criminal might use when committing a crime. He may walk into a room, see someone counting money, and think: "I need money, so I will take his. Since the man is protecting the money, I will get rid of him. I'll shoot him. He'll drop to the floor. I will then take the money and run. I'll leave by the window."

On the other hand, the conscious mind, being objective and self-aware, is able to step outside itself. It can reason both deductively and inductively. To reason inductively is to move backward from result to cause. A police detective, for example, would arrive at the scene and begin reasoning backward in an attempt to figure out how the crime was committed and who might have done it.

The fact is that the subconscious mind is entirely under the control of the conscious (objective) mind. As a result, the subconscious will work diligently to support or to bring into reality whatever the conscious mind believes and feels is true. And let me say here that it is important that the two things—belief and feeling—be in concert. The subconscious may be even more responsive to what you feel about something than what you actually think about it. So do not just hope what you want in life will come about. Feel it already is. If you really feel that something must be, then it must be. At least, that is how the subconscious interprets it, and so the subconscious mind will go to work to bring it into reality.

Something else to know is that the subconscious mind does not understand, or perhaps it simply doesn't hear the words "no" and "not." Suppose, for example, you're a tennis player. You're in a big match, it's close, and you arrive at a crucial point. Your opponent is going to serve next, and as you pass by him at the net when changing ends, you say, "You're playing great today, Henry. Don't blow it. This is a big point coming up. Whatever you do, don't double fault."

You've started Henry worrying, and on top of that, his subconscious mind doesn't hear or understand the word "don't." All it hears is "double fault," and it takes that as a directive. Try as he might to do otherwise, Henry will double fault.

Actually, I advise you not to play such a dirty trick. As the saying goes, "What goes around, comes around," and you don't want that sort of negative behavior coming back at you. More will be written about this. The important thing to remember is that self-talk and coaching should always be framed in a positive way. Do not say to yourself, "don't double fault." Rather, you should phrase the desired outcome in a positive way, such as, "Make this one an ace!" Thoughts put into words are powerful. Positive affirming self-talk will change your life. Say to yourself and think: "I'm good, I'm wealthy, I'm the luckiest person I know!"

Think Positive and Rid Yourself of Fear

Let's dig into the issue of fear because fears are thoughts, beliefs, and feelings, and thoughts, beliefs, and feelings are what create your reality.

To learn what you fear, tune into your moment-to-moment stream of consciousness and observe what makes you worried, anxious, resentful, uptight, afraid, angry, and so on. Step outside yourself and identify unsettled emotions, tugs and urges that have become part of your programming. Slow down and consider what triggered a negative emotion. Did your temper flare? Why? Why was it so important for things to go a certain way? If you trace what you felt back to its cause, in most cases, you will come to a particular variety of fear, and it's been said that only two fears are instinctive: the fear of falling and loud noises. Other fears were acquired, and whatever was acquired can be disposed of.

According to some experts, the fears that hold people back can be grouped under one of six headings:

1. the fear of poverty (or failure),
2. the fear of criticism,
3. the fear of ill health,
4. the fear of the loss of love,

5. the fear of old age,
6. and the fear of death.

I will focus on the fear of poverty (failure) because it is the biggest deterrent many face when it comes to accumulating and achieving wealth and success. Like anything that's held in the mind long enough, it is self-fulfilling—a major reason being that traits develop that bring it about. For example, are you a procrastinator? An underlying fear of failure is probably the root cause and can be counted upon to produce the unwanted result.

Are you overly cautious? Do you see the negative side of every circumstance or stall for the "right time" before taking action? Do you worry that things will not work out, have doubts (generally expressed by excuses or apologies about why you probably won't be able to perform), do you suffer from indecision (which leads to someone else, or circumstances, making the decision for you)?

Are you indifferent? This generally shows itself as laziness or a lack of initiative, enthusiasm or self-control.

Step back and listen for internal voices that say "can't" or "don't" or "won't" or "too risky" or "why bother?"

How do you get rid of them? Shoo them away.

No matter who you are—president, king, or knave—the only thing over which you have absolute control is your thoughts. As was stated earlier, you may not be able to control what thoughts enter your mind, but you can decide whether to discard one or to keep it. Thoughts are not part of you, and that means you can decide that a thought is

Proof We're All One

counterproductive and throw it away, or you can turn it over and over in your mind, in effect nurture it, and let it grow. You are who you are and what you are, and what you will become, because of your thoughts. Make them productive.

If negative thinking is a problem for you, go to YouTube and find hypnotic meditations to listen to that will plant positive thoughts in your mind in place of negative ones. Play them over and over for at least a month. Get all that junk out of your head. One I recommend is "Money & Success - Bedtime Guided Meditation" by Barry Tesar. You can find it on YouTube.

I suspect you have heard about "The Power of Positive Thinking," that positive thoughts are much more likely to produce good results than negative thoughts. I'm reminded of The Little Engine That Could, an American fairy tale published numerous times in illustrated children's books and movies since its original debut in 1930. The Little Engine was a railroad locomotive that was tasked with pulling a long, heavy train—one that seemed much too large for it—up and over a mountain. But even so, the Little Engine was determined and kept telling itself over and over, "I think I can, I think I can." It was a struggle, but the Little Engine persevered and finally succeeded.

It's a good story and a valuable one to teach young children the benefits of optimism and hard work. The problem is, many of us today were not taught that lesson as children, and in fact, feelings of frustration, discontent, and dissatisfaction were ways of solving problems that many of us "learned" as infants. For example, if a baby is hungry, he or she expresses discontent by crying. Lo and behold, a warm and tender hand appears magically out of nowhere and brings a bottle of milk. Later on, if the baby is uncomfortable, again, he or she will again express dissatisfaction, and the same warm, comforting hands magically appear and solve the problem. That's fine for babies but, unfortunately, many children continue to get their way and have their problems solved by indulgent parents merely by continuing to express their feelings of frustration when things don't work the way they want. All they have to do is feel frustrated and dissatisfied, express their dissatisfaction, and the problem will be solved. Sometimes what have become known as "helicopter parents" continue to cater to their children in this way all the way through high school, college, and beyond.

This way of life "works" for infants, and for some children. But it does not work in adult life when a person is out in the world on his or her own. Yet many continue to expect, perhaps unconsciously, that it will work. They seem to think that by feeling discontented and expressing their grievances—if only they feel put upon enough—life, or someone will take pity on them, rush to their aid and solve their problems. Let me assure you that 99.99 percent of the time that is not going to happen. It is my advice that you take responsibility for every aspect of your life.

Imagine, for example, you are lucky enough to land an entry-level job as a management trainee in a big corporation. With you in training are several other bright young men and women fresh out of business school. Imagine the way things work in this company is often not to your liking. Management trainees, for example, are relegated to cubicles with five-foot-high walls affording little or no privacy, while the senior staff all have corner offices with large windows and spectacular views of the East River. You spend a good deal of time grumbling to yourself and to others about this injustice, subconsciously believing that will get you out of that cubicle and into a corner suite. Your fellow trainees, on the other hand, spend their time making positive suggestions and anticipating and providing for the needs of customers as well as for fellow workers higher up on the corporate ladder. Whom do you suppose is most likely to be first to break out of his or her cubicle? The one who constantly complained? Or the one that consistently delivered the goods?

Don't you feel a twinge inside that intuitively "knows" the positive attitude, the attitude of service to others, will inevitably win the day? That "twinge" is a message from something inside you that knows the correct answer called "intuition."

If you have been ignoring that feeling when it comes, now is the time for you to begin recognizing such messages. They have a light and airy feeling to them, even though they may seem to run counter to egocentric notions, such as, "The first order of business is to look out for number one." That egocentric notion may work in the short term, but in the long term, it is bad advice. The fact is that it's always best to under-promise and over-deliver to customers and bosses—as well as to anyone else for that matter. By over-delivering, your reputation grows as you create positive vibes and positive opinions of you by those with whom you come in contact. A reputation that you are someone who can be counted on can only lead to good outcomes and opportunities for you in the long run.

Let's consider for a moment why some people may spend their valuable time on earth grumbling and complaining away opportunities to get ahead. It's often because they have felt frustrated and defeated for so long—ever since they were babies in a crib and while growing up with indulgent parents—that those feelings have become ingrained. Their minds are in a kind of holding pattern, and it's never occurred to them to step outside of themselves in order to get in touch with intuition that would tell them, if they would only listen, that grumbling and complaining are counterproductive and accomplish nothing. Until they wise up, they will continue—to their own detriment—to radiate those feelings, and as sure as night follows day, their discontent will lead to failure.

No matter what your mindset, if you want to change it, it's important to know that thoughts and feelings are intertwined. It might be said that feelings are the soil in which thoughts and ideas grow. If you are habitually grumpy and in a bad mood, you need to lighten up and begin seeing the glass half full. Moreover, when you begin working toward a goal, try thinking how you will feel when you reach it—and then actually make yourself feel that way. I'm serious. Conjure up the feeling of "Success!" The thrill of accomplishment will communicate the belief to your subconscious mind that it's inevitable you are going to achieve what you have set out to accomplish. The feeling creates the belief, and the belief creates the feeling. A mental model of success will be etched into your subconscious, and that the desired outcome will surely come about.

Let's say you are pursuing a challenge and fervently want to accomplish it. Assuming you have the education, the knowledge, and the qualifications required to reach your goal, and assuming you feel strongly about it at an emotional level, you will almost certainly realize success. This will happen because your subconscious mind, which is connected to the Infinite Mind and all other minds, will go to work and act like a magnet, drawing to you what you need. The greater your desire, the more powerfully your subconscious will mind strive to produce results.

In summary, belief and emotions are the keys. It's important to feel the joy of having accomplished what you set out to accomplish before it actually happens. This will convince your subconscious the goal has already been reached and the Infinite Mind will cause it to be reached as a result. As Jesus said in Mark 11:24 (NLT), "Therefore I tell you, whatever you ask for in prayer, believe that you have received it, and it will be yours." Jesus might have added, "And because you believe you have already received it, feel the joy of having received it, too."

Be Likable and Appealing to Others

Becoming successful will require effort and work, and that means things will be easier if you have help along the way. The most likely source of that help will be friends, partners, and mentors. Obviously, the going will be easier and you will attract more help if people like you and want to work with you. Therefore, it should go without saying that it's important to be someone others want to be around—someone people would like and want as a friend. That means you need to be someone who "talks the talk," and "walks the walk." Perhaps you know a person who does the opposite. If not, you are likely to come across someone like that in your business dealings, so be prepared and never, ever be one of them. In public such people talk openly—some even brag and boast—about the importance of having integrity and doing the right thing. But in private it's a different story. They bad mouth people and do things that aren't consistent with the honest-John public persona they hope to project. People quickly see through these phonies. As the old saying goes, "Say what you mean, and mean what you say," and people will respect you for doing so.

This is not to say you shouldn't be competitive. You definitely should have a competitive attitude, and of course you should enjoy the thrill of victory. It's important to have a winning attitude, winning surely beats the alternative, and very important, people want to go with a winner—they want to "hitch their wagon to a star." But it's also important to understand that you should not gloat about winning. Be humble, and realize that often a surprisingly big payoff will come in the form of goodwill if you are a gracious loser. It is far better to give credit to others, especially where credit is due. Unless your name is

Muhammad Ali [1942-2016]—the champion boxer known for his colorful bragging and boasting—never brag or boast. Such behavior indicates immaturity and insecurity.

Obey the Laws of Physics

I'm sure you have heard, "What goes around comes around." It's true and something you always ought to keep in mind. In the East, it's called "Karma." In reality, it is simply "cause and effect," a law of physics—Newton's Third Law of Motion: "For every action, there is an equal and opposite reaction." Throw a rock into a pond, and you will disturb the harmony of the pond. You were the cause, and the effect was the splash. The ripples flow out, and they flow back until harmony is restored. In the same way, disharmonious actions by a person go out into the world and come back upon back upon that person until harmony is restored. That is why it is always best to "live and let live," and to follow the Golden Rule, which everyone knows is to "do unto others as you would have others do unto you."

If you are a Materialist and think matter is all that exists—that there is no God and nothing spiritual—you might come to the conclusion you can do whatever you want, harm whomever you want, and never have to suffer any consequences. But that is not how things work. I agree it may work for a while, but eventually you'll get back what you gave out because just as there are laws of physics, there are laws of metaphysics. People must obey them, and nations must as well. Take Japan and Germany in World War

Proof We're All One

II as examples. Both countries had astonishing victories in the beginning. Each country benefited from a zeitgeist that they were invincible. But they ended up being crushed because the atrocities they committed came back upon them with a vengeance.

Let's dig a little deeper into the topic of Karma. If "what goes around comes around," the more you give, the more you will receive. In other words, what you send out into the world will return to you. Therefore, the more you assist others, in the long run, the more assistance will come to you. And it's important to do it not for your own gain but because it is who you are. It is what is right. It is not some theoretical, do-gooder idea. It will work in your day-to-day life, provided your motive is to help others without seeking or expecting anything in return because, as has been said, nobody likes a phony—so don't be someone who is hoping to score brownie points in order to receive something in exchange. Be your best, unselfish self!

All well and good, you say, but what if such benevolence is not in my nature? Here's my advice: Change your nature. Absolutely do not expect anything in return and force yourself to "fake it until you feel it." As you do good deeds and see how your actions affect others, you will eventually come to enjoy doing them. In time, you will want to help others because it will make you feel good to do so. Get to know yourself. The nice thing about being human is we get to choose the person we want to be. Choose characteristics that make you proud, proud of everything about you.

Proof We're All One

Let's say you are in business and someone lashes out at you in anger. What should you do? Say, for example, you're the owner of a business and a competitor or vendor files a lawsuit against your business based on some real or perceived grievance. Rather than strike back and escalate the problem, the first thing to do is to gather the facts. Get everyone from your company together who knows anything about the vendor and what may actually have happened to cause the dispute. The group should do its best to determine what really happened and come up with all of the issues and interests the other side may have.

Such questions ought to be answered such as, "What can we do to get them what they want or need?" To persuade them to withdraw the suit, you might decide to offer to them a contract or an attractive benefit in some other area of the business or part of the country. Your objective is to restore harmony rather than provoke and escalate chaos. Finding a win-win solution often can do just that. You and your group will probably want to brainstorm and evaluate the effects different actions will have on the specific individuals bringing suit. What does the leader of the other side want, personally?

The bottom line is that you can get into a back-and-forth fight that might end up harming your business as well as the other guy's, or instead, by making a peace offering, you might be able to mitigate the situation. In most cases, if you are gracious toward others, you will defuse what could turn into a much worse situation. It has been said that if you want to make a friend, you must be a friend.

Of course there are unreasonable individuals in this world, and there are bullies who refuse to be placated. Sometimes people will sue for no good reason at all, except that they want to get something for nothing. Often such individuals understand only one thing, and that's a punch in the nose—figuratively speaking, of course. If that's the case, do what you have to do, but only after an attempt at reasoning with them. The figurative punch in the nose should be used only as a last resort. The point is that things are more likely to fall into place for you if you are able to establish and maintain harmony with others. And, happily, the benefits do not stop with simply falling into place.

Always Keep Your Life in Balance

You are probably familiar with the ancient Chinese symbol composed of a white "yin" interlocking a black "yang" that represents the dual nature of things. It symbolizes that we live in a world that is composed of opposites: Up, down, black, white, good and evil. Without the tension opposites create, nothing would or could exist—everything would fall apart. Follow the advice of this book and attain total success but do not allow complacency to set in. Always seek new challenges, realizing that without one, self-destruction may be the result. By keeping success and challenge in balance, it will be possible to maintain your position and retain your success.

It can also be comforting to know there can be no growth without at least some discontent. Deep within, you know what is best for you. There is an urge built into you that pushes you to strive for growth, and for most of us, growth will not continue without some agitation and discontent. So study your dissatisfactions. They will tell you what you are about to leave behind and possibly point you in a new direction. Be willing to be uncomfortable. It is the only way you grow.

As you contemplate your future course, it is also important to realize you can only attract that which you feel worthy of. Self-esteem is critical to success. The truth is you are not what you have, and you are not simply what you do. Beneath your fear programming, you and every other human being on earth is a magnificent creature, a spiritual being encased in a vehicle consisting of billions of living, conscious cells. To repeat what was written earlier, you are the pinnacle of life on Earth. Fear and negative programming are the primary obstacles with the power to prevent you from realizing your full potential. The more you can let go of fears, the more you program your mind to think positive thoughts, the higher your self esteem will be, and the more options you will have, and the more risks you will be willing to take. The more you like yourself, the more others will like you, and the more worthy you will feel. This is true for everyone, and it is what needs to be taught from birth, and it is what needs to be taught and reinforced in school.

You can have anything you want if you can give up the belief you cannot have it and replace it with the belief you can. Of course, you must get the education necessary and learn the skills you need to create what you want. "Where your attention goes, your energy flows." You attract what you are and that which you concentrate upon. If you are negative, you draw in and experience negativity. If you have a loving attitude toward others, you draw in and experience love. You can attract to yourself only those qualities you possess. So, if you want peace and harmony in your life, you must become peaceful and harmonious. If you want wealth, you must develop a wealthy attitude.

The mind is engaged in an endless state of growth and reorganization. This is good because it means you can discard old, negative-thought baggage and replace it with productive thoughts and ideas. You might think of yourself as the pilot of a speedboat. The boat is your mind, and you can steer it in any direction you want. The wake behind the boat is your past, and good or bad, there's no reason to look back because you are headed for a bright future ahead. It is absolutely true that it is possible to reprogram yourself. It can be done, as mentioned previously, by repeatedly listening to success-meditation recordings, or with visualization techniques. If you feel anxiety in crowds, for example, imagine yourself feeling relaxed while in a crowd of people. If you are nervous about public speaking, imagine yourself calmly addressing a crowd of adoring fans that want nothing more than to hear what you have to say. Practice the skill over and over until you create a new self-image for yourself.

Above All, Take Time to Make a Life

Perhaps you have heard the saying, "Never become so busy making a living that you do not have time to make a life." That's advice worth taking that should start with doing what's necessary to maintain a healthy balance between your work life and your home life. If you are married, for example, pay attention to your spouse. Keep the romance that brought you together going strong. If you have children, spend time with each one of them. Listen to them. Take them to lunch or to an amusement park. Ask them questions. Coach their little league teams. Constantly give them encouragement. Attend their recitals. Applaud their successes. The fact is you have within you everything required for your life to become a virtual paradise if you choose to accept that it is possible, if you have the desire to make it so, and if you maintain a healthy balance in all areas of your life. We live in a world of abundance, although many populating our planet appear to view it as a universe of scarcity. Too bad for them, but it does not have to be so for you.

What else might you do to maintain balance and thereby create a happier life? Here are a few suggestions:

- Take five minutes, twice a day to affirm your goals, dreams, and desires. Most of us do not achieve our goals, not because we are too lazy or not talented enough, but because we forget about them and focus our efforts elsewhere.
- Spend some time in nature. Even if it is just ten minutes a day, take the time to go for a short walk or sit in a place surrounded by the natural world. Release the stress of the day by com-

muning with Mother Nature, and you will soon feel recharged.

- Exercise. Your body is your temple and your most important possession. Take care of it. If it is not in top form, neither will you be in top form. Exercising, eating healthy and taking care of your amazing vehicle is vitally important for you to be able to produce at the highest levels.
- Take time to meditate. The biggest improvements in our lives almost always come from within. An effective way to release the limiting beliefs and destructive thoughts that may plague you is to meditate for fifteen to thirty minutes a day. Regular practice of meditation has been proven scientifically to change your brain chemistry, lower blood pressure, help you sleep better, feel less stressed, and more.
- Smile a lot. A smile can change the world. Not only for you but also for the people with whom you interact. Practice a genuine smile and give joy to the world. Impact the world today and every day by smiling at everyone that passes by.
- Find more ways to have fun. Life does not have to be a strict, gloomy experience we struggle through. Instead it can be full of amazing twists and turns. Think of it as an adventure because that's what it is. Approach it as such.
- Make sure you laugh out loud at least once a day. Do something childish, or completely

weird. Be yourself, have fun, and laugh at your own jokes.
- Remember that everything begins as a thought or idea. Ideas and experiences create beliefs that in turn, create your reality. If you are not satisfied with your current reality, you must change your beliefs and your behavior. Beliefs should be changed when you realize which ones are not working for you. Change that belief, and your life will change.

I'll finish with this thought. To change for the better, you must first recognize the destructive or disharmonious thinking and behavior you need to eliminate. Understand that you don't have to change how you feel about it, but rather, you simply need to change what you are doing. The Buddha was right when he said, "It is your resistance to what is that causes your suffering." By suffering, he meant everything that doesn't work in your life. This might include relationship problems, loss of loved ones, loneliness, sickness, accidents, guilt, financial hardship, unfulfilled desires, and so on. When you accept what you cannot change, you will be in position to set that aside and stop worrying about it. And yes, you absolutely should change what you are able to change that needs changing—no doubt about it. But you also need to have the wisdom to accept what you cannot change. Out of acceptance will come detachment, which will enable you to enjoy the positive aspects of life without being distracted by the negative. Why waste energy focusing on things from the past when you can move on and put those things behind you? You are the pilot of the physical vehicle you inhabit and your life.

I'm reminded of something my mom told me many times when I was growing up, "If at first you don't succeed, try, try again." That phrase became imprinted on my brain, and I'm glad it did because it is one of the most important keys to success. Thomas Edison, for example, conducted 10,000 experiments before he found a way to make an incandescent light bulb that actually worked for a substantial length of time. When asked how it felt to fail 10,000 times, he replied, "I didn't fail. I found 10,000 ways not to make a light bulb." For him, every failure was a small success bringing him closer to accomplishing his goal. The same can be true for you if you decide to look at life that way. Decide now to take the attitude that you and you alone are responsible for everything that happens to you, and take full responsibility for every aspect of your life.

Your attitude toward life and your experiences are returned to you as love and joy, or as confusion, trouble, and heartbreaking experiences. The way to mitigate the punishments and maximize the rewards is to take total responsibility for your life, grow in wisdom with each new life lesson, and seek harmony in everything you do.

# # #

About Stephen Hawley Martin

Stephen Hawley Martin is a former marketing executive and consultant and the author of more than three-dozen books, including five published novels, half a dozen business management titles, and quite a few self-help books and metaphysical investigations. He is a former principal of the world-renowned advertising agency, The Martin Agency, the firm that created the GEICO Gecko and "Virginia is for lovers." Today, Stephen is editor and publisher of The Oaklea Press. Listed in *Who's Who in America,* and best known as a award-winning author, Steve is the only three-time winner of the *Writer's Digest* Book Award, having won twice for fiction and once for nonfiction. He has also won First Prize for Visionary Fiction from *Independent Publisher* and First Prize for Nonfiction from *USA Book News.* He and his wife of 35 years live in central Virginia.

To get in touch with Stephen and learn about other books he has written, visit his website:

www.shmartin.com

This two-book volume contains the bestselling title, *Life After Death, Powerful Evidence You Will Never Die* and the sequel, *Heaven, Hell & You.* As one reviewer, a medical doctor, wrote: "Extraordinary findings . . . will keep readers on the edge of their seats as they burn through this well written book's pages."

Kindle: ASIN: B07J46QQW8
Paperback: ISBN-10: 1727782038

If you believe that soul growth is the overarching reason we incarnate on Earth, this book was written for you. In it Stephen Hawley Martin describes in detail an incredible, mystical experience he had, and he shares insights he brought back with him that explain how you can raise your vibration to the next level in order to achieve joy in this life and bliss in the next.

Kindle: ASIN: B09GHY3SRN
PB: ISBN 979-8479020360

If you believe you are, or that you might be a spiritual being having a physical experience and wonder why you're here on Earth, this book provides the answers. Read it with an open mind. If you take to heart and follow the guidance it gives, your suffering might not vanish entirely, but you will be better able to deal with the little that remains.

Kindle: ASIN: B08CYBNZX4
PB: ISBN: 979-8666306222

You may believe humans are spiritual beings having a physical experience, but are you sure why we're here and what we ought to do about it? This book will tell this you this and much, much more because, as the record shows, the accuracy of information revealed by Edgar Cayce's more than 14,000 psychic readings was nothing less than extraordinary.

Kindle: ASIN: B07L7GF3HH
Paperback: ISBN-10: 1790978114

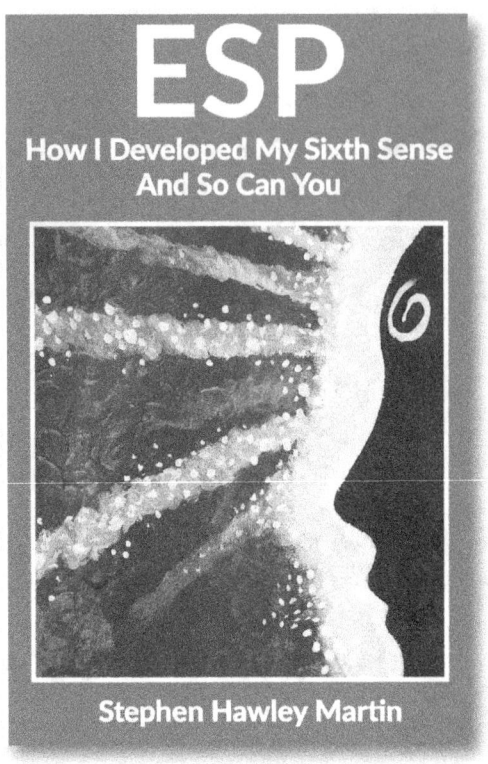

All the knowledge of the universe resides within you because at a deep level all minds, past and present, are connected. Everything that has ever happened, every thought, every idea is there. The trick is to draw out information when you need it. In this book Stephen explains how he learned to do so and how you can, too.

Kindle: ASIN: B07HHFFWP8
Paperback: ISBN-10: 1723835250

This 5-Star rated, fast-paced thriller is based on the true nature of reality as revealed in the book you hold in your hands. The heroine travels to the Caribbean Island of Martinque to save her father and learns the secret of life in the process. A page-turner, this novel won First Place for Fiction from *Writer's Digest* and First Place for Visionary Fiction from *Independent Publisher* magazine.

Kindle: ASIN: B08S7MG4WM
Paperback: ASIN: B08SB6VG9L

www.ingramcontent.com/pod-product-compliance
Lightning Source LLC
Chambersburg PA
CBHW072101110526
44590CB00018B/3270